D0792779

Planet in Peril

An Atlas of Current Threats to People and the Environment

Editorial management
Alain Gresh, Philippe Rekacewicz and Dominique Vidal (journalists at *Le Monde diplomatique*, Paris)

Jean Radvanyi (professor at the National Institute of Oriental Languages and Civilisation, Paris)

Catherine Samary (lecturer in economics at Paris-Dauphine university)

Cartography
Philippe Rekacewicz in collaboration with Emmanuelle Bournay, Laura Margueritte and Cécile Marin

Graphic design and layout
Boris Séméniako, with Nathalie Le Dréau

Cover design
Boris Séméniako

Documentation
Olivier Pironet, Clea Chakraverty, Nicolas Harvey, Laurent Morin and Ophélie Wiel

GRID-Arendal contributors to the English edition
Janet Fernandez Skaalvik in collaboration with Karen Landmark, Ieva Rucevska, Otto Simonett and Thore-André Thorsen

English translation
Harry Forster

This atlas is the fruit of discussions between a group of people on key environmental issues. It owes a great deal to the imagination, knowledge and creativity of its various contributors.

Published by UNEP/GRID-Arendal, Arendal Norway
and *Le Monde diplomatique*, Paris France 2006.

Foreword

Otto Simonett

Planet in Peril: an Atlas of Current Threats to People and the Environment uses text, maps and diagrams to illustrate the short and long term relationship between the world's population and its ecosystems and natural resources. It brings together a wealth of information from the most up-to-date sources.

This atlas – a translation of the environment pages of *L'Atlas 2006 du Monde diplomatique* – is the result of long-standing cooperation between *Le Monde diplomatique* and GRID-Arendal. In our main area of work – providing environmental information for decision-making – the media are the most important vehicle for reaching our target audiences. Over the last few years we have intensified our cooperation with journalists and the media involving them in UNEP and GRID-Arendal activities in the field and engaging seasoned environmental journalists in our training activities. This has resulted in better environmental reporting and active environmental media networks in places such as Central Asia, the Caucasus and Africa.

There is an environmental dimension to most of the world's crises and disasters, whether they concern the global energy situation or highly localised conflicts. It is consequently obvious that environmental issues cannot be taken out of their geo-political and socio-economic context. This holistic, well researched view of today's global issues and its presentation in a concise, understandable and very visual form is the main strength of *Le Monde diplomatique's* atlas.

With this publication of priority environmental topics we hope to foster greater awareness, explain connections and contribute to solutions.

Contents

Warning: this planet is fragile and its inhabitants are vulnerable

Introduction

Philippe Rekacewicz

In March 2005 the United Nations published the results of the biggest ever study of the world's ecosystems[1]. For five years 1,300 people, representing 95 different countries and a wide range of cultural and scientific horizons, contributed to this large-scale assessment. It demanded thorough cross-checking and involved more than 30 regional studies, raising a series of issues that demanded clear conclusions. Much as the title of the Assessment Board's statement (Living Beyond Our Means), its findings are unequivocal. Over the last 50 years mankind has used and transformed ecosystems more than in any equivalent period in history. We have irreversibly damaged more than half the planet's ecosystems. All the observations registered during the study show that biodiversity is declining.

With the acceleration in population growth that started in the 1950s, pressure on our natural environment has increased apace. Our growing demands for raw materials, food, fresh water and energy just must be satisfied, but in so doing we have substantially reduced nature's ability to continue providing the services we need in our daily lives. In the meantime our standard of living has improved considerably. But how long will this paradox hold true? A wood stove will go on radiating heat long after we stop putting logs in it. It continues to provide a service (heat), giving us the impression it is still working, long after it has in fact gone out. It is quite possible that nature, totally exhausted in areas where our demands have been greatest, will no longer – perhaps very soon – be able to provide the services we need to stay alive. Such an eventuality jeopardises the survival of the poorest, most vulnerable members of the world's population. The standard of living is still improving all over the world, but very unevenly. Some 2.7 billion people (40% of the world's population) live in a state of poverty or extreme poverty (with less than $2 a day)[2] and another 850 million often go hungry[3].

Over the last few decades the extraction and transport of raw materials to major industrial centres, subsequent processing (in heavy industry, manufacturing, and the fabrication of hi-tech equipment), and the conveyance of final products to market have placed enormous pressure on the environment and on people. What impact are these activities having on the environment?

The following pages try to provide an answer to this basic question in brief but comprehensive terms. The world is experiencing very rapid economic growth, but mankind must cope with several major challenges. We must manage the supply of fresh water, raw materials and energy resources. We must mitigate the consequences of climate change and find viable solutions for neutralising or destroying several billion tonnes of hazardous (chemical, or civil and military nuclear) industrial waste. And we must find ways of ensuring every single one of the world's 6.4 billion people gets enough to eat and proper health care[4].

Understanding our troubled world is not an easy task and the pace of change is ever quickening. For example in 2005 the Intergovernmental Panel on Climate Change (IPCC)[5] and the authors of the Arctic Climate Impact Assessment (ACIA)[6] observed that warming was progressing much faster in the Arctic regions than predicted in even the worst-case scenario of the 2001 IPCC report[7]. For many years chemical firms in rich countries were in the habit of relocating activities to poor countries (such as India, Brazil, Thailand and Mexico) where laws and regulations were considered more accommodating. These countries are now introducing stricter rules. India, for instance, has just refused the French aircraft carrier Clémenceau access to its ship-breaking yards, arguing that it contains asbestos (from 50 to 500 tonnes, depending on sources) and constitutes a hazard for Indian workers[8]. In southern Africa, where the HIV-Aids pandemic is a source of increasing concern, life expectancy in Zambia has dropped from 51 to 32 years in less than 15 years[9], returning to the same level as in the 1940s. The "big" issues that fill the headlines generally make us forget that atmospheric pollution, essentially industrial in origin,

is a persistent problem in rich (for instance, southeast Germany, southern Poland and the Czech Republic) and poor countries (Asia's seasonal "brown cloud"). However, the respiratory diseases it causes mainly affect the poor.

Thorough investigation has shown that two of the world's worst industrial accidents did not happen by chance. They were the result of human carelessness. The Bhopal disaster, in December 1984, was caused by the failure of safety systems that had not been properly maintained for months, as part of a drive to cut production costs[10]. The poisonous cloud that escaped from the plant, and subsequent pollution, killed several thousand people and injured more than 500,000 others. The explosion of a nuclear reactor at Chernobyl, in April 1986, was the catastrophic result of an experiment designed to test a new system for boosting electricity output. There is still disagreement about the damage it caused, ranging from 32 dead to several tens of thousands exposed to severe radiation, with vast areas of land in Europe contaminated and unusable[11]. But in neither case, was the disaster inevitable. On the contrary both accidents were due to foolhardy attempts to cut production costs. At some point in the industrial chain of command a single person or group of individuals opted to disregard safety rules and give priority to productivity at the expense of human life.

Nor were Bhopal and Chernobyl isolated cases. Seveso, Minimata, the Prestige, the Exxon Valdez, Semipalatinsk and Mururoa are all names that will remain in our memories as major failures, perhaps even crimes, committed by our industrial society which, at every step in its development, has proved incapable of protecting the environment it exploits and the people dependent upon it.

1. Millennium Ecosystem Assessment (http://www.millenniumassessment.org).
2. "The dimension of poverty", in The Wealth of the Poor, Managing Ecosystems to Fight Poverty, World Resources Institute (WRI), Washington, 2005.
3. Food and Agriculture Organisation of the United Nations (FAO).
4. See L'Atlas du Monde diplomatique, edited by Alain Gresh, Jean Radvanyi, Philippe Rekacewicz, Catherine Samary and Dominique Vidal, special issue, Paris, February 2006.
5. http://www.ipcc.ch.
6. Arctic Climate Impact Assessment (ACIA), Impact of a Warming Arctic, AMAP, CAFF and IASC, Cambridge University Press, 2004.
7. Climate change 2001: synthesis report, contribution of working groups I, II, and III to the IPCC third assessment report, Geneva.
8. Agence France Presse, 13 February 2006
9. Population, Development and HIV/Aids With Particular Emphasis on Poverty: The Concise Report, United Nations, Population Division, 2005; UNAIDS/UNICEF, 2004 (2003 figures).
10. Bhopal: India's man-made disaster, Olivier Bailly, Le Monde diplomatique, English language edition, December 2004.
11. Guillaume Grandazzi and Frédérick Lemarchand, Les silences de Tchernobyl: l'avenir contaminé, Editions Autrement, Paris, 2004.

Polar ice caps melting

Global warming is not affecting the planet evenly and most of the existing models forecast that it will be greater in the northern hemisphere. With an overall increase of 2°C, temperatures in the Arctic could increase by a factor of two or three. The southern hemisphere, would also be affected, though less severely.

The North Pole is already showing signs of substantial change. Over and above considerable variations between seasons and years, the surface area of the ice pack has diminished by 10% in 30 years. By the end of the 21st century half of it may have disappeared.

Some try to look on the bright side, highlighting the opening of new sea passages for trade and easier access to oil and gas fields in the far north of America and Siberia, which contain 40% of global reserves. But the disadvantages largely outweigh such benefits. The most serious immediate problem concerns the Gulf Stream.

Preliminary research has revealed that the strength of the current dropped by 20% between 1950 and 2000. Paradoxically it could temporarily result in much colder weather in Europe.

Worse still melting of the ice caps could increase the pace of global warming, by reducing refraction of solar radiation – 80% on ice, compared with 30% on bare earth and 7% on the sea. In some places the permafrost (permanently frozen ground) is melting. Not only does it support buildings and infrastructure, but it also contains very large quantities of methane gas. The Arctic Council, whose members include the United States, Canada and Russia, has failed so far to do anything to counter these risks.

Melting of the Arctic ice in itself does nothing to raise the level of the oceans, as the ice is already floating on the sea. But gradual melting of the Greenland ice sheet and glaciers in other parts of the world could make a significant difference. Measurements by

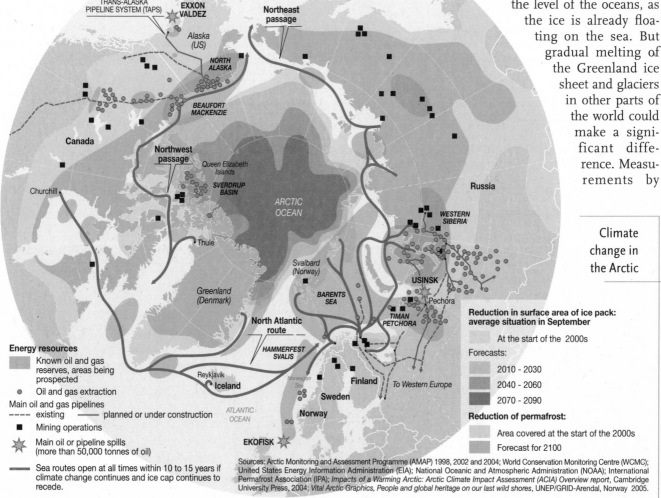

Energy resources

- Known oil and gas reserves, areas being prospected
- Oil and gas extraction

Main oil and gas pipelines
- - - - existing ——— planned or under construction
- Mining operations
✶ Main oil or pipeline spills (more than 50,000 tonnes of oil)

—— Sea routes open at all times within 10 to 15 years if climate change continues and ice cap continues to recede.

Climate change in the Arctic

Reduction in surface area of ice pack: average situation in September

- At the start of the 2000s

Forecasts:
- 2010 – 2030
- 2040 – 2060
- 2070 – 2090

Reduction of permafrost:
- Area covered at the start of the 2000s
- Forecast for 2100

Sources: Arctic Monitoring and Assessment Programme (AMAP) 1998, 2002 and 2004; World Conservation Monitoring Centre (WCMC); United States Energy Information Administration (EIA); National Oceanic and Atmospheric Administration (NOAA); International Permafrost Association (IPA); *Impacts of a Warming Arctic: Arctic Climate Impact Assessment (ACIA) Overview report*, Cambridge University Press, 2004; *Vital Arctic Graphics, People and global heritage on our last wild shores*, UNEP/GRID-Arendal, Norway 2005.

faster

the Topex-Poseidon satellite currently indicate a 2.4 millimetre annual rise in sea level. That would result in a rise of at least 25 centimetres by the start of the next century, but increasing numbers of scenarios are forecasting a rise of one or several metres, if melting of certain parts of the Antarctic is confirmed. Setting aside such uncertainty, it appears that a third of the rise is caused by dilatation of the sea water due to the temperature increase. Melting glaciers account for a further third. As for the remainder, recent studies suggest that melted ice from the South Pole could already be accounting for as much as 15% of the total rise.

RISING SEA LEVEL

Until very recently scientists thought only the Antarctic peninsula was affected. It warmed up 3°C between 1974 and 2000 and it was here that the huge Larsen ice shelf broke free in 2002. If all the ice on the peninsula melted the sea level would rise by an additional 45 centimetres. However it is not directly connected to the southern polar ice cap which, until recently, was thought to be stable and unlikely to be affected by global warming for at least a century. Then, in October 2004, NASA revealed that the temperature of

some parts of the continent might increase by more than 3.6°C by 2050. In December 2004 a team belonging to the British Antarctic Survey observed that the western part of Antarctic was losing 250 cubic kilometres of ice a year. It remains a relatively small amount, but if the rate of loss increased, water from this area could ultimately raise the sea level by 8 metres. For the time being only Eastern Antarctic, much the largest part (equivalent in ice to a 64-metre rise in sea level), appears to have been spared.

In addition a reduction in the Antarctic ice pack could have a disastrous effect on aquatic wildlife. In particular krill, a tiny shrimp that lives on seaweed growing under the ice, play a key role in the marine food chain, feeding squid, fish and cetaceans. Krill stocks appear to have dropped by 80% over the last 30 years. Combined with overfishing worldwide and increasing damage to the coral reefs, this undoubtedly constitutes an additional source

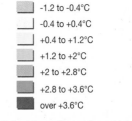

Antarctic temperature rise by 2050

Temperature variations

▨	-1.2 to -0.4°C
☐	-0.4 to +0.4°C
▨	+0.4 to +1.2°C
▨	+1.2 to +2°C
▨	+2 to +2.8°C
▨	+2.8 to +3.6°C
▨	over +3.6°C

Source: National Aeronautics and Space Administration (NASA), 2004. Based on a map drawn up by Frédéric Durand.

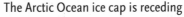

The Arctic Ocean ice cap is receding

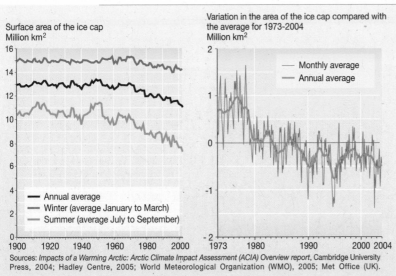

Surface area of the ice cap
Million km²

— Annual average
— Winter (average January to March)
— Summer (average July to September)

Variation in the area of the ice cap compared with the average for 1973-2004
Million km²

— Monthly average
— Annual average

Sources: *Impacts of a Warming Arctic: Arctic Climate Impact Assessment (ACIA) Overview report*, Cambridge University Press, 2004; Hadley Centre, 2005; World Meteorological Organization (WMO), 2005; Met Office (UK).

■ On the web

> Arctic Climate Impact Assessment (ACIA):
www.acia.uaf.edu

> International Arctic Science Committee (IASC):
www.iasc.no

> Arctic Council:
www.arctic-council.org

> Center for International Climate and Environmental Research-Oslo (CICERO):
www.cicero.uio.no

> Laboratoire d'océanographie dynamique et de climatologie (LODYC):
www.lodyc.jussieu.fr

> World Meteorological Organization (WMO):
www.wmo.ch

> International Conference on Arctic Research Planning (ICARP):
www.icarp.dk

Point of no return for

The Kyoto protocol came into force on 16 February 2005, heralding the advent of a more mature attitude. Mankind, we were told, had finally woken up to the increasing pressure that it is exerting on the environment. Unfortunately a closer look shows that such claims have more to do with wishful thinking than actual fact.

Forecasts of global warming have become more alarmist in recent years. The 2001 report by the Intergovernmental Panel on Climate Change (IPCC) confirmed that the greenhouse effect had significantly increased since the 19th century. Carbon dioxide (CO_2) emissions contributed to a worldwide temperature increase of 0.8°C between 1860 and 2000. The same report predicted that temperatures would rise faster, increasing by 1.4°C to 5.8°C between 2000 and 2100. Given that during the last ice age, 15,000 years ago, the planet as a whole was only about 5°C colder, this would be a considerable increase.

A study published by Oxford University in 2005, based on the results of 2,578 computer simulations, forecast an even higher temperature rise: between 1.9°C and 11.5°C, most of the results ranging from 2°C to 8°C. The greatest source of concern is the notion of the point of no-return. Due to climatic inertia, even if drastic measures were taken now, the impacts of the current disturbance would persist for years. They might even be irreversible. A consensus has emerged that the critical threshold could correspond to an overall temperature rise of 2°C. To prevent this, the CO_2 concentration should not exceed 550 parts per million (ppm), or perhaps even 400 ppm. But in fact it rose from 270 ppm around 1850 to 380 ppm in 2004, an unprecedented increase in the 420,000 years of climate history that scientists have been able to reconstitute. Over that period the CO_2 concentration varied between 180 ppm and 280 ppm. The current annual rate of increase stands at 2 ppm, which means a critical threshold could be reached within 10 to 30 years. It also means we need a fourfold cut in CO_2 emissions by industrialised countries by 2050.

THE WEIGHT OF EVIDENCE

Admittedly we are dealing with forecasts, not absolute certainties. But the importance of the risks and the growing consensus among scientists should encourage us to apply the precautionary principle and take effective measures. What, then, would the Kyoto protocol achieve if it was fully implemented, in other words if the United States ratified it and Europe met its commitments? It would only reduce global warming forecast for the end of the century by 0.06°C (or 2% to 3%). Furthermore the protocol does not set

Beyond the critical threshold

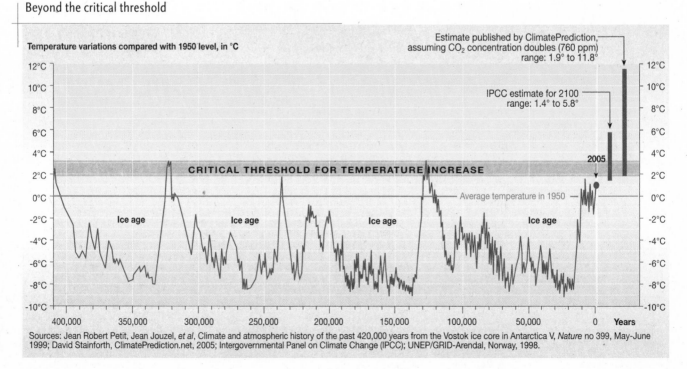

Temperature variations compared with 1950 level, in °C

Estimate published by ClimatePrediction, assuming CO_2 concentration doubles (760 ppm) range: 1.9° to 11.8°

IPCC estimate for 2100 range: 1.4° to 5.8°

2005

CRITICAL THRESHOLD FOR TEMPERATURE INCREASE

Average temperature in 1950

Ice age   Ice age   Ice age   Ice age

400,000   350,000   300,000   250,000   200,000   150,000   100,000   50,000   0   Years

Sources: Jean Robert Petit, Jean Jouzel, *et al*, Climate and atmospheric history of the past 420,000 years from the Vostok ice core in Antarctica V, *Nature* no 399, May-June 1999; David Stainforth, ClimatePrediction.net, 2005; Intergovernmental Panel on Climate Change (IPCC); UNEP/GRID-Arendal, Norway, 1998.

global warming

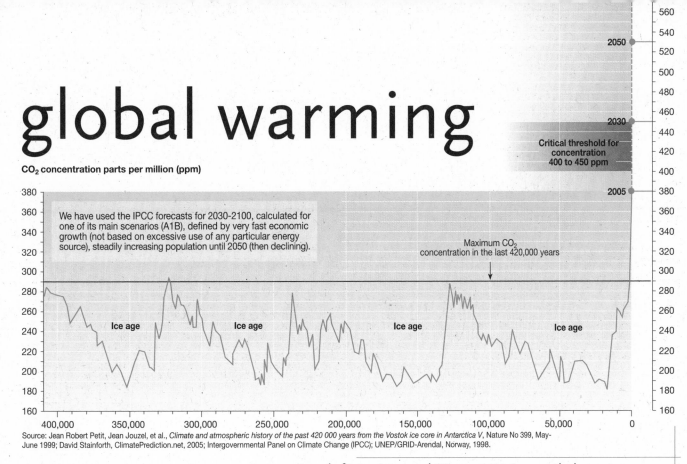

CO₂ concentration parts per million (ppm)

We have used the IPCC forecasts for 2030-2100, calculated for one of its main scenarios (A1B), defined by very fast economic growth (not based on excessive use of any particular energy source), steadily increasing population until 2050 (then declining).

Maximum CO₂ concentration in the last 420,000 years

2050

2030

Critical threshold for concentration 400 to 450 ppm

2005

Ice age   Ice age   Ice age   Ice age

Source: Jean Robert Petit, Jean Jouzel, et al., *Climate and atmospheric history of the past 420 000 years from the Vostok ice core in Antarctica V*, Nature No 399, May-June 1999; David Stainforth, ClimatePrediction.net, 2005; Intergovernmental Panel on Climate Change (IPCC); UNEP/GRID-Arendal, Norway, 1998.

Record of temperature and CO₂ concentration over the last 400,000 years

any limits on emissions in developing countries, which understandably want to catch up with industrialised countries. The failure, at the end of 2004, of the negotiations at the Buenos Aires conference, which was supposed to prepare a follow-up to Kyto, is an indication of the present deadlock.

Yet, although the forecasts are still uncertain, the signs of an imminent upset are accumulating. The last decade (1995-2004) was the hottest since the start of regular records in the 19th century. It saw an increase in the number of extreme events: the frequency and intensity of El Nino increased; the heat wave that affected Europe in 2003 could become a recurrent feature; in 2004 the US and Asia suffered an unprecedented number of typhoons. It is perhaps too soon to say they are all connected, but the available evidence increasingly points that way.

Several structural phenomena have been confirmed, even if it is still difficult to predict their consequences accurately. In addition to warming in the polar regions (see section on pages 8-9), the increase in the temperature of the oceans is damaging coral reefs, a habitat essential to marine wildlife. The sea level could rise by between 25 centimetres and 1 metre due to dilatation of water as it warms up. Nor does that allow for melting of the ice caps. Some studies are predicting

150 million climate refugees by 2050. Changes in rainfall patterns could affect farming and the areas in which diseases propagate. The consequences for biodiversity are also likely to be particularly serious, with many species struggling to adapt to such rapid changes. Even without climate change human beings have already caused the sixth largest wave of biological extinction the Earth has ever known, simply on account of the destruction and pollution we habitually wreak.

On the web

> United Nations Framework Convention on Climate Change (UNFCCC):
www.unfccc.int

> Intergovernmental Panel on Climate Change (IPCC):
www.ipcc.ch

> Worldwatch Institute:
www.worldwatch.org

> Global Resource Information Database (GRID-Arendal):
www.grida.no/climate

Average temperature variation on Earth since 1861

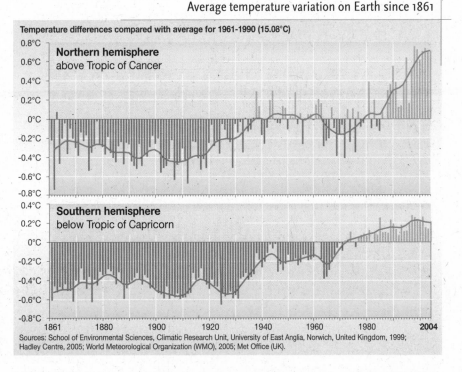

Temperature differences compared with average for 1961-1990 (15.08°C)

Northern hemisphere above Tropic of Cancer

Southern hemisphere below Tropic of Capricorn

Sources: School of Environmental Sciences, Climatic Research Unit, University of East Anglia, Norwich, United Kingdom, 1999; Hadley Centre, 2005; World Meteorological Organization (WMO), 2005; Met Office (UK).

Water becoming a rarity

Despite the international community's commitments many people still do not enjoy the right of access to clean water and half the world's population is in danger of running short of this vital commodity in 30 years.

More than 1.1 billion human beings do not have access to drinking water and 2.4 billion lack proper sanitary facilities. For some people water may seem abundant, but reserves are very unevenly spread. Whereas a few countries hold 60% of the planet's fresh water reserves, Asia, home to 60% of the world's population, only has 30% of the total. Water shortages are a permanent state of affairs in a triangle stretching from Tunisia down to Sudan and across to Pakistan. Each person has an average of less than 1,000 cubic metres of fresh water a year, a situation described as a "chronic shortage".

Water quality is also a problem. The larger the amount consumed, the more waste water is produced. In developing countries 90% of waste water and 70% of industrial waste runs straight into the surface water without any form of treatment.

As a result more than 5 million people die every year of water-related diseases, 10 times more than the number of victims of armed conflicts. The world's population is set to rise from 6 billion people in 2000 to 8 billion in 2025. The average amount of fresh water available per person per year will consequently decrease by almost a third. If water use goes on increasing at the present rate the UN estimates that in 20 years' time 1.8 billion people will be living in areas affected by a constant water shortage, with 5 billion others located in places where it will be difficult fully to satisfy their needs.

As the population drift from the countryside to the towns continues the situation will deteriorate further, with increasing numbers packing into the planet's giant metropoles. By

Water usage

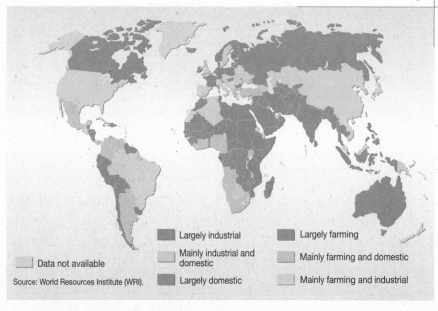

▣ Largely industrial		▣ Largely farming	
▣ Mainly industrial and domestic		▣ Mainly farming and domestic	
▣ Data not available	▣ Largely domestic		▣ Mainly farming and industrial

Source: World Resources Institute (WRI).

Urban development changing the picture

In the beginning was a village ...

■ Water is drawn from just below the surface of the groundwater (blue arrows).

■ Rainfall filters into the ground, circulating horizontally and replenishing groundwater (green arrows).

■ Waste water is partly removed by the sewage system, but also filters into the ground (brown arrows).

... which soon became a town

■ The level of the groundwater drops significantly, so wells must be bored increasingly deep.

■ Large quantities of waste water pour into the ground, contaminating the surface water (brown areas).

■ Subsidence may occur due to the sudden drop in the level of the groundwater, leaving a weakened substratum in which empty pockets replace the water.

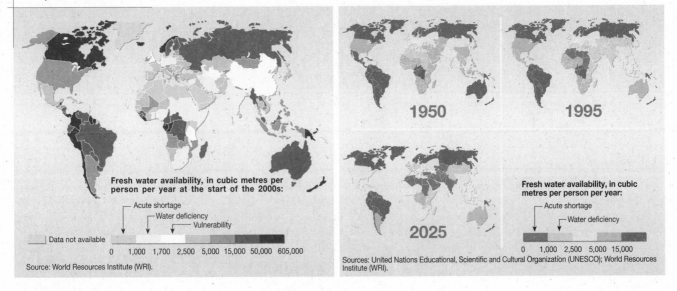

Fresh water availability, in cubic metres per person per year at the start of the 2000s:

Acute shortage
Water deficiency
Vulnerability

Data not available

0   1,000 1,700 2,500 5,000 15,000 50,000 605,000

Source: World Resources Institute (WRI).

1950

1995

2025

Fresh water availability, in cubic metres per person per year:

Acute shortage
Water deficiency

0   1,000 2,500 5,000 15,000

Sources: United Nations Educational, Scientific and Cultural Organization (UNESCO); World Resources Institute (WRI).

2020 27 of the world's 33 largest cities (population exceeding 8 million people) will be located in the South. The corresponding influx of people will lead to a 40% increase in domestic water consumption.

But wastage increases as the standard of living improves. The many amenities appearing in well-off homes encourage extravagant use of water, regardless of its relative scarcity and its rising cost (which, driven upwards by private utilities, may be prohibitive for the poor). Europeans currently use eight times more fresh water on a daily basis than their grandparents. The average inhabitant of Sydney, Australia, uses more than 1,000 litres of drinking water a day, compared with 300 to 400 litres for an American and 100 to 200 litres for a European. In some developing countries the average daily consumption per capita barely exceeds a few litres.

Vast amounts of water are simply wasted. Only 55% of all water produced is actually used. The rest is lost, either

because it drains away or evaporates during irrigation, or because it leaks from the mains. To feed the world's population the productivity of farming must substantially improve. Irrigation, which already accounts for 70% of all the water produced, will need to increase by 17% over the next 20 years.

Attempts to solve the water shortage based exclusively on technology, such as desalination of sea water, will

only have a limited impact due to their cost. We must improve the efficiency of our water usage, particularly for irrigation, refurbish drinking water production and distribution resources, protect reserves and combat pollution. According to various funding agencies this will require an annual investment of $180bn over the next 25 years, compared with $75bn at present.

Unfortunately there is disagreement as to which remedies should be promoted. Privatisation of water, recommended by international donors and some governments, still only concerns 5% of global resources. Many non-governmental organisations condemn this mercantile approach, maintaining that access to water is a "basic human right", that should either be free or charged at its real cost. But even then the poorest people will not unable to pay for their water. We consequently face a dual challenge: we must manage water wisely and protect the right of access of the poorest people to this vital resource.

■On the web

> **International Rivers Network (IRN):**
www.irn.org

> **United Nations Educational, Scientific and Cultural Organisation (UNESCO):**
www.unesco.org/water/

> **The Water Barons:**
www.icij.org/water

> **Planète bleue:**
www.planetebleue.info

> **The World Conservation Union (IUCN):**
www.iucn.org/themes/wani

> **H²O:** www.h2o.net

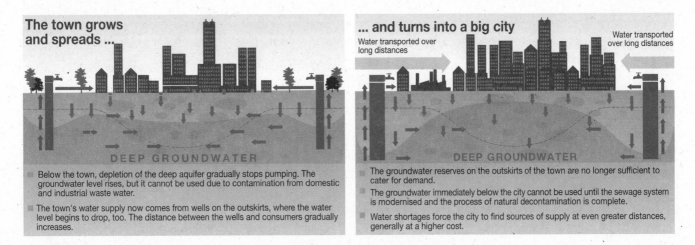

The town grows and spreads ...

DEEP GROUNDWATER

■ Below the town, depletion of the deep aquifer gradually stops pumping. The groundwater level rises, but it cannot be used due to contamination from domestic and industrial waste water.

■ The town's water supply now comes from wells on the outskirts, where the water level begins to drop, too. The distance between the wells and consumers gradually increases.

... and turns into a big city

Water transported over long distances

Water transported over long distances

DEEP GROUNDWATER

■ The groundwater reserves on the outskirts of the town are no longer sufficient to cater for demand.

■ The groundwater immediately below the city cannot be used until the sewage system is modernised and the process of natural decontamination is complete.

■ Water shortages force the city to find sources of supply at even greater distances, generally at a higher cost.

Ocean resources under threat

The oceans supply about 80% of all living aquatic resources, amounting to 110m tonnes. (Mt). The rest (28 Mt) comes from inland waters. At sea, production relies to a large extent (80%) on fishing, simply harvesting natural resources, the remainder coming from mariculture, which encompasses the various techniques of fish farming.

For thousands of years fishing was relatively inefficient, but the situation changed radically over the last century, thanks to major advances in the techniques used to catch and store fish. Catches totalled 20 Mt in 1950, rising to 70 Mt in 1970 then stabilising between 80 Mt and 90 Mt. The spectacular increase in 1950-70 was largely due to the development of industrial uses for fish, transforming it into by-products (meal and oil) for use manufacturing pet food.

This market engulfs huge volumes of fish (sometimes as much as 35% to 40% of catches). It has caused overfishing of certain species and major crises, such as the massive drop in

The planet's one ocean – for the various oceans form a single ecosystem – covers 361m square kilometres, or 71% of the Earth's surface. Exploitation of renewable and non-renewable resources has steadily increased. Some renewable resources are the focus of keen rivalry. No sooner do we realise their potential than they threatened by over-exploitation.

herring catches in the north-east Atlantic in 1968, or a similar fall in anchovy catches off the coast of Peru from 1972 onwards. These crises led to the setting up of exclusive fishing grounds extending 200 nautical miles out from the coastline. Within each area the relevant country enjoys exclusive fishing rights and can apply quotas for specific species. As marine wildlife lives mainly on the edges of the oceans, it must of necessity be shared between neighbouring countries, resulting in disputes such as the cod war that flared between Iceland and the United Kingdom in 1975. Norway and Russia have still not managed to reach agreement on fishing limits. In Asia, overfishing is one of the reasons for the boom in fish farming, with annual production rising from 6 Mt to 25 Mt in just 25 years.

The availability of fish as a foodstuff (with a global average of about 16 kg per person per year) is stable but very unevenly spread. China, where consumption is expanding fast, and the developed countries enjoy plentiful supplies, in contrast to countries in Africa and Central America, already suffering from chronic malnutrition.

Other uses for the sea are being explored, in particular scope for gener-

Fishing yields

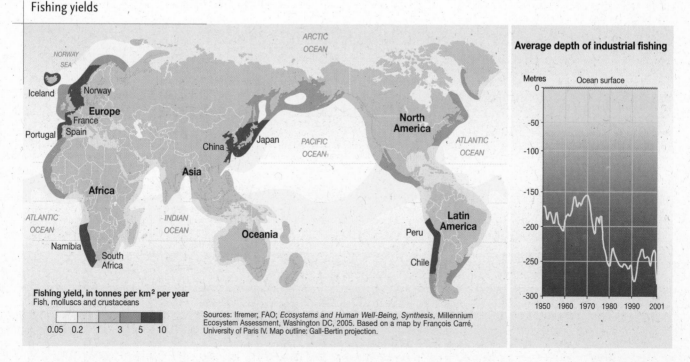

Fishing yield, in tonnes per km² per year
Fish, molluscs and crustaceans

0.05 0.2 1 3 5 10

Average depth of industrial fishing

Sources: Ifremer; FAO; *Ecosystems and Human Well-Being, Synthesis*, Millennium Ecosystem Assessment, Washington DC, 2005. Based on a map by François Carré, University of Paris IV. Map outline: Gall-Bertin projection.

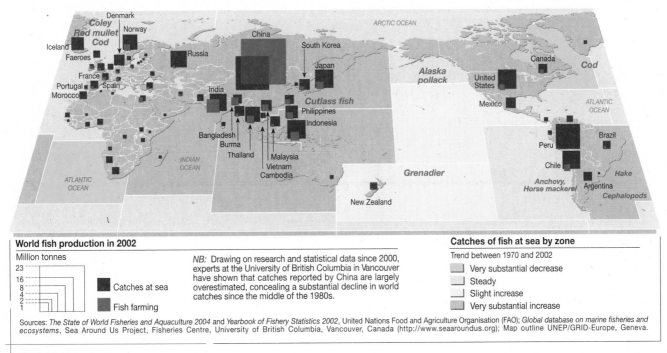

World fish production in 2002

World fish production in 2002

Million tonnes

23
16
8
4
2
1

■ Catches at sea

■ Fish farming

NB: Drawing on research and statistical data since 2000, experts at the University of British Columbia in Vancouver have shown that catches reported by China are largely overestimated, concealing a substantial decline in world catches since the middle of the 1980s.

Catches of fish at sea by zone

Trend between 1970 and 2002

☐ Very substantial decrease
☐ Steady
☐ Slight increase
☐ Very substantial increase

Sources: *The State of World Fisheries and Aquaculture 2004* and *Yearbook of Fishery Statistics 2002*, United Nations Food and Agriculture Organisation (FAO); *Global database on marine fisheries and ecosystems*, Sea Around Us Project, Fisheries Centre, University of British Columbia, Vancouver, Canada (http://www.seaaroundus.org); Map outline UNEP/GRID-Europe, Geneva.

ating energy from the movement of the water (waves, swell and currents), or from the vertical temperature gradient between warm surface water and the chill ocean depths. Although there is huge potential, attempts to use such energy sources have so far only been experimental and limited in scale. Pilot projects include the tidal power station on the Rance estuary in northern France, built in 1966, and a similar facility in northern Russia, built two years later.

Non-renewable resources found in and under the sea comprise mainly fossil fuels such as coal, with coalfields, mined on land, extending out into the sea, and above all hydrocarbons, currently the focus of active prospecting. But the seabed conceals other mineral resources too.

UNEXPLOITED RICHES

Most of the oil and gas under the seafloor is exploited on the continental shelves, at depths not exceeding 200 metres. But the rising price of crude oil makes it likely that deep-sea reserves, at depths of 1,500 to 3,000 metres, will be prospected, thus prolonging exploitation of oilfields previously thought to be nearing depletion.

Minerals and ore are also to be found on the seabed, but they are still little used. They include ore containing iron and sulphur, placers (alluvial deposits rich in metals and gems), sedimentary materials used in construction (sand, gravel and pebbles), and phosphorite rocks from which phosphates can be extracted. In 1970-

80 the nodules containing various metals scattered all over the deep seabed attracted considerable interest, but the cost of bringing them to the surface was prohibitive. The same is true of the metal-rich muds deep in the Red Sea.

Lastly seawater itself provides sodium chloride, on salt marshes, magnesium and bromine, accounting for 80% of the world's needs. And of course, after desalination, it is a source of fresh water.

■ On the web

> United Nations Food and Agriculture Organisation (FAO):
www.fao.org/fi/

> Intergovernmental Oceanographic Commission (IOC):
www.ioc.unesco.org

> International Council for the Exploration of the Sea:
www.ices.dk

> Institut français de recherche pour l'exploitation de la mer (Ifremer):
www.ifremer.fr

> Onefish:
www.onefish.org

> International Maritime Organisation (IMO):
www.imo.org

Dwindling stocks of Atlantic cod

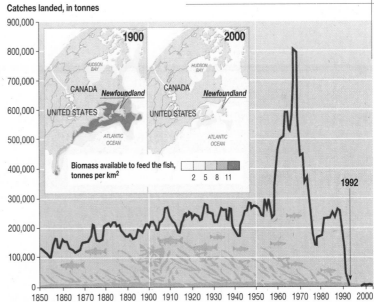

Catches landed, in tonnes

Sources: *Ecosystems and Human Well-Being, Synthesis*, Millennium Ecosystem Assessment, Washington DC, 2005; *Global Database on Marine Fisheries and Ecosystems*, Sea Around Us Project, Fisheries Centre, University of British Columbia, Vancouver, Canada (http://www.seaaroundus.org).

Nuclear power for civilian

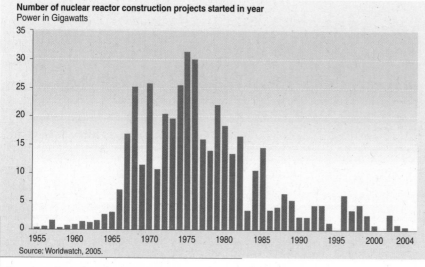

Number of nuclear reactor construction projects started in year
Power in Gigawatts

Source: Worldwatch, 2005.

Construction of nuclear reactors

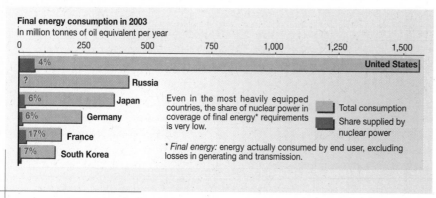

Final energy consumption in 2003
In million tonnes of oil equivalent per year

4% — United States
? — Russia
6% — Japan
6% — Germany
17% — France
7% — South Korea

Even in the most heavily equipped countries, the share of nuclear power in coverage of final energy* requirements is very low.

Total consumption
Share supplied by nuclear power

* *Final energy*: energy actually consumed by end user, excluding losses in generating and transmission.

Nuclear power, a backup solution

Nuclear power only makes a minor contribution to world energy consumption. Given the average age (about 22 years) of the nuclear reactors still in service and nuclear power stations' limited share (barely 2%) of the market for new electricity production facilities, the situation is unlikely to change in the immediate future. In the meantime there is still no solution to the problem of nuclear waste and the risk of proliferation.

On 26 April 1986 the explosion at Chernobyl nuclear power station sent a cloud of radioactive dust round the world. More than 400,000 people were permanently evacuated from contaminated areas. Many countries placed restrictions on farming, slaughtered livestock and destroyed crops. In 2005 there were still restrictions on 379 farms and more than 74,000 hectares of pasture in the United Kingdom, 2,500 kilometres from the scene of the disaster.

One of the most striking features of the event is that such a small amount of material could have generated so much power and caused such widespread damage. The explosion at Chernobyl released less than 27 kilograms of cesium-137, but it resulted in planetary contamination, accounting for three-

quarters of the overall pollution. The spent-fuel pools at La Hague, in northwest France, contain about 300 times as much cesium. At Tokaimura in Japan the accidental fission of 1 milligram of uranium killed two people, after causing them terrible suffering, and irradiated several hundred others living in the vicinity. On 9 August 1945 the explosion of a bomb containing about 1 kilogram of plutonium, in the air above Nagasaki, killed 74,000 people instantly and injured at least as many, not to mention long-term effects.

Although the military were quick to show an interest in nuclear power, civil stocks now represent the largest accumulation of radioactive material. The strategic potential of civil nuclear power facilities, facilitated by the dissemination of technical know-how, has focused attention on attempts by countries such as Iran and North Korea to develop their own nuclear programmes.

PROLIFERATION

Nuclear power plays a relatively small and gradually decreasing role in global energy. Taking into account losses during electrical power production and transmission, nuclear power barely covers 2% of the world's energy requirements. Some 440 reactors, located in 31 countries, supply 16% of global commercial consumption of electricity and 6% of primary energy. The six main producers – United States, France, Japan, Germany, Russia and South Korea – generate three-quarters of all nuclear electrical power. France, the outstanding exception (nuclear power stations produce 75% of its electricity) accounts for 45% of all nuclear power generation in the European Union.

Unless major technical advances are made the situation seems unlikely to change. Even if the service life of reactors is extended to 40 years, it will be necessary, if only to maintain the existing installed capacity, to commission about 80 reactors over the next 10 years (equivalent to a reactor every six weeks), adding a further 200 over the following 10 years. According to the International Atomic Energy Agency

and military use

(IAEA) there were only 24 nuclear power stations under construction in May 2005.

Some nuclear materials, in particular highly enriched plutonium and uranium, may be used for civil purposes or in explosive devices. Attempts to distinguish between civil and military uses make increasingly little sense

■ On the web

> Centre de documentation et
de recherche sur la paix et les conflits (CDRPC):
www.obsarm.org

> Federation of American Scientists (FAS):
www.fas.org/nuke/

> Arms Control Association (ACA):
www.armscontrol.org

> Power Reactor Information System (PRIS):
www.iaea.org/programmes/a2/

> Plutonium Investigation (WISE-Paris):
www.wise-paris.org

technically and often provide an excuse for disregarding measures to control proliferation. In all the countries possessing nuclear weapons, progress in civil nuclear science has benefited arms development, and vice versa. Although civil nuclear power plays a relatively minor role in energy production, the strategic potential of the materials involved and the inherent risk of a military or terrorist attack have steadily increased. The stock of "civil" plutonium exceeds 230 tonnes worldwide and it is increasing. It represents at least twice the amount contained in the 30,000 nuclear warheads thought to exist.

The Nuclear Non-Proliferation Treaty calls on its signatories (China, the US, France, the UK and Russia, the acknowledged nuclear-weapon states) to negotiate "general and complete disarmament". In practice they

◎ Explosions in civil nuclear facilities during Soviet period

Source: Arctic Monitoring and Assessment Programme (AMAP), 1998.

Serving the mining industry

have never stopped developing new weapons. The US and Russia have substantially reduced the number of deployed warheads, but most of these weapons were considered obsolete. A genuine initiative for disarmament would involve resumption of negotiations for a treaty banning the production of enriched plutonium and uranium.

The two sides to an industry

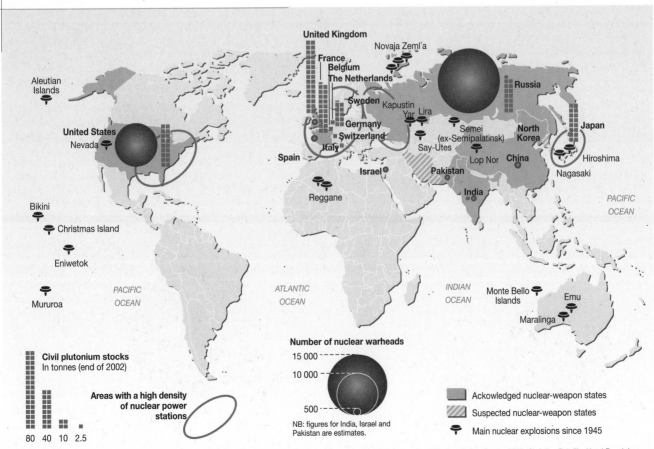

Civil plutonium stocks
In tonnes (end of 2002)

80 40 10 2.5

Areas with a high density of nuclear power stations

Number of nuclear warheads

15 000
10 000
500

NB: figures for India, Israel and Pakistan are estimates.

Acknowledged nuclear-weapon states

Suspected nuclear-weapon states

Main nuclear explosions since 1945

Sources: International Atomic Energy Agency (IAEA), Vienna; Carnegie Endowment for International Peace, 2005; International Nuclear Safety Centre, 2002; Christian Bataille, Henri Revol, Les incidences environnementales et sanitaires des essais nucléaires effectués par la France entre 1960 et 1996 et éléments de comparaison avec les essais des autres puissances nucléaires, French National Assembly (report no 3571) and Senate (report no 207), Paris, 2002.

Renewable energy, fact and fiction

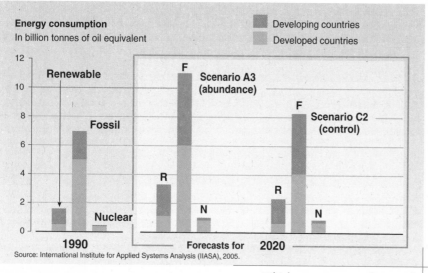

Energy consumption
In billion tonnes of oil equivalent

Developing countries
Developed countries

Renewable

F — Scenario A3 (abundance)

Fossil

F — Scenario C2 (control)

Nuclear

R N R N

1990   Forecasts for   2020

Source: International Institute for Applied Systems Analysis (IIASA), 2005.

Which energy source for 2020?

Renewable energy technologies have made considerable progress. Windmills and solar panels, modern wood-burning boilers, biofuels, bioclimatic buildings are all widely available, often at competitive prices.

Almost all the scenarios advanced by energy specialists include very encouraging forecasts for renewable energies, ranging from 2,500m tonnes to 3,300m tonnes of oil equivalent (toe) in 2020, much higher than oil at present. It is the case, in particular, for the scenarios published by the International Institute for Applied System Analysis, an authority in the matter. But its forecasts are based on the assumption that use of renewable energy sources by developing countries (760m toe) will be three times higher than in developed countries (175m toe, or barely 20% of the potential resources). This disparity is particularly striking because it is much easier to promote the use of renewable energy in rich rather than poor countries. They may replace fossil fuels already in use, catering for existing, solvent demand, whereas in the South their successful introduction depends on there being additional, solvent demand.

The example of solar, or photovoltaic (PV), energy is particularly instructive. There has been much talk of off-grid solar panels, hailed as a miracle solution for 2 billion people in developing countries without electricity. Over the last 20 years, at considerable cost in aid, solar panels have brought electrical lighting and a limited power supply to 500,000 people.

But the 1,999.5 million others are still without electricity. Even with a hundredfold increase in the rate of installation it would take at least 400 years to equip them all. Given that off-grid PV electricity costs three to five times more than its diesel-powered equivalent and that the panel itself only represents 20% of the total, it is clear that solar panels are unlikely to become competitive even in the medium term. Unless, of course, the price of crude oil reaches $150 or $200 a barrel, which would dash any hopes of development in poor countries anyway. In short, setting aside heavily subsidised schemes, there is no viable market for off-grid solar panels. The same applies to any gain associated with a reduction in carbon dioxide emissions which at best would only cover 20% of the investment.

Electricity production in the world

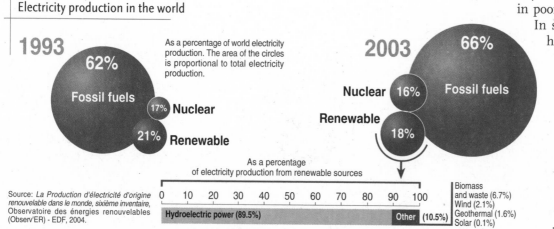

1993

As a percentage of world electricity production. The area of the circles is proportional to total electricity production.

62% Fossil fuels

17% Nuclear

21% Renewable

2003

66% Fossil fuels

Nuclear 16%

Renewable 18%

As a percentage of electricity production from renewable sources

Source: *La Production d'électricité d'origine renouvelable dans le monde, sixième inventaire,* Observatoire des énergies renouvelables (Observ'ER) - EDF, 2004.

0 10 20 30 40 50 60 70 80 90 100

Hydroelectric power (89.5%)   Other (10.5%)

Biomass and waste (6.7%)
Wind (2.1%)
Geothermal (1.6%)
Solar (0.1%)

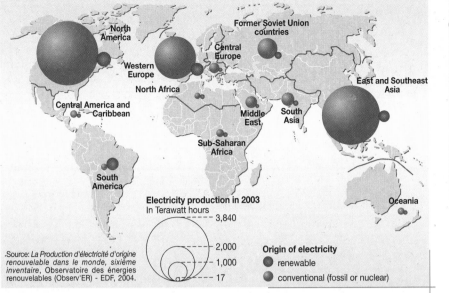

Electricity production in 2003
In Terawatt hours

3,840
2,000
1,000
17

Origin of electricity
- renewable
- conventional (fossil or nuclear)

Source: *La Production d'électricité d'origine renouvelable dans le monde, sixième inventaire*, Observatoire des énergies renouvelables (Observ'ER) - EDF, 2004.

Electricity output from renewable energy sources still marginal

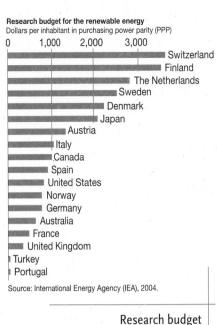

Research budget for the renewable energy
Dollars per inhabitant in purchasing power parity (PPP)

0   1,000   2,000   3,000

Switzerland
Finland
The Netherlands
Sweden
Denmark
Japan
Austria
Italy
Canada
Spain
United States
Norway
Germany
Australia
France
United Kingdom
Turkey
Portugal

Source: International Energy Agency (IEA), 2004.

Research budget

It is consequently unrealistic to claim that off-grid solar panels will save the world from under-development. It is already possible to supply the necessary energy straight away and more cheaply, but in other ways. On the other hand it is obviously tempting for firms and governments in industrialised countries, with the blessing of public opinion, to use development grants to boost funding for research in this field.

It looks very much as if countries in the North, concerned about greenhouse gas emissions, are advocating massive use of renewable energies everywhere but at home, despite the fact that the main markets are located there, along with the essential financial, technical and industrial resources. If we really want renewable energies to achieve their full potential, we must make several changes:

■ saving energy must become a genuine priority. Otherwise, if energy consumption continues to increase, no production technique, sustainable or otherwise, will be efficient enough

in the immediate future to prevent a climatic disaster;

■ rich countries must finally make up their minds to capitalise on the considerable potential they have in this field, leaving oil, at its current price, to the developing countries rather than imposing policies that are often ill suited to their short-term needs. Some countries, such as Germany, have already

got the message, launching large-scale, grid-connected wind or solar-powered schemes. Others, particularly France, are doing all they can to resist the trend;

■ we must help countries in the South with significant biomass, hydraulic or solar-power potential to concentrate their own R&D resources on projects that use such potential wisely and capitalise on it locally.

World potential for 2020

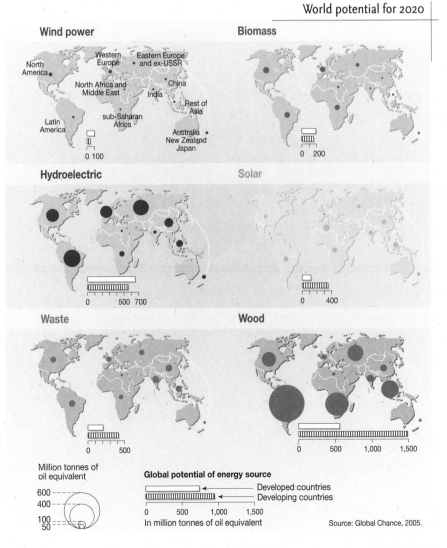

Wind power

Biomass

0 100

0 200

Hydroelectric

Solar

0   500 700

0   400

Waste

Wood

0   500

0   500   1,000   1,500

Million tonnes of oil equivalent

600
400
100
50

Global potential of energy source
- Developed countries
- Developing countries

0   500   1,000   1,500
In million tonnes of oil equivalent

Source: Global Chance, 2005.

■ **On the web**

> **International Institute for Applied Systems Analysis (IIASA):**
www.iiasa.ac.at

> **World Energy Council (WEC):**
www.worldenergy.org

> **Les cahiers de Global Chance:**
www.agora21.org/edition21.html

> **Observatoire des énergies renouvelables:**
www.energies-renouvelables.org

> **French national energy debate:**
www.debat-energie.gouv.fr

Planet in peril

Weapons for rich ...

Weapons of mass destruction (WMD) have only one thing in common, their potential for killing large numbers of people. The term covers nuclear, chemical and biological weapons, as well as ballistic missiles, their main vector. On the sidelines dirty bombs belong to the arsenal of terrorism.

The term "weapons of mass destruction" (WMD) surfaced during the American presidential election campaign in 1996. Prior to that they had been referred to as nuclear, biological and chemical (NBC) weapons. Setting aside the lethal capacity they all share, they differ largely by their means of production and use. Developing nuclear weapons, the WMD par excellence, is a state monopoly, whereas individuals or small groups can manufacture chemical and biological weapons.

Several groups of countries currently possess nuclear weapons. The

Countries suspected of developing biological weapons	Syria Egypt	**United States** **Russia** **Israel** **Iran** **China** **North Korea**	**Countries suspected of developing chemical weapons**
		India Pakistan France South Africa United Kingdom	
		Countries with a nuclear weapons programme	

first category comprises the five acknowledged nuclear-weapon states: the US, France, China, Russia and the UK. Apart from the US attacks on Hiroshima and Nagasaki in 1945, they have never used their weapons, except to test them (more than 2,000 nuclear tests have been carried out since 1945, 530 in the atmosphere and underwater, and about 1,500 underground). The trend among this group is towards partial disarmament (there were about 16,500 nuclear warheads worldwide in 2005, compared with almost 70,000 in 1985, at the peak of the cold war) but new developments in the US and Russia may reverse the trend.

With the Nuclear Non-Proliferation Treaty (NPT), which came into force in 1970, these countries unsuccessfully attempted to block the spread of nuclear weapons. India and Pakistan joined the group of acknowledged nuclear-weapon states in 1998, but without signing the NPT. Other "threshold states" are thought to have

secretly developed nuclear weapons. Israel, which started its military programme in 1957 after the Suez crisis, has probably made the most progress. North Korea, which has withdrawn from the NPT, claims to possess several nuclear devices. Iran will soon be able to produce nuclear weapons. Despite pressure from Europe and the US it is reluctant to shelve its plans, arguing that it is surrounded by hostile powers. Iraq no longer counts as a threshold state, an independent US commission having concluded that it no longer had any stocks of biological and chemical weapons and that its nuclear programme was "inoperative", invalidating two of the justifications for the preventive attack in March 2003.

It is quite possible for small countries to develop biological and chemical weapons. Referred to as the "poor man's" WMD, some are relatively cheap and easily obtained. A distinction should nevertheless be

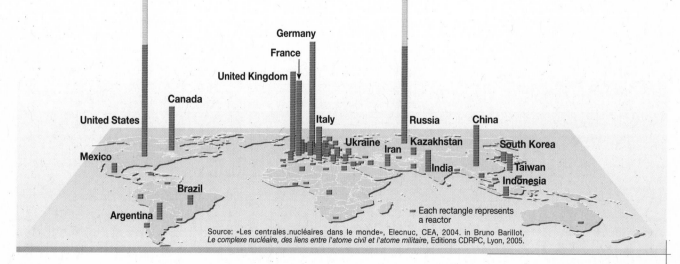

Source: «Les centrales nucléaires dans le monde», Elecnuc, CEA, 2004. in Bruno Barillot, *Le complexe nucléaire, des liens entre l'atome civil et l'atome militaire*, Editions CDRPC, Lyon, 2005.

World's research reactors at start of 2000s

and poor

drawn between military weapons, which require large-scale industrial facilities, and toxic agents that can be synthesised in small quantities in an ordinary laboratory.

Over the last 15 years western countries, apart from the US, have started reducing the size of their chemical and biological stockpile. But at the same time some developing countries have started upgrading their weapons, increasing their strategic value. Egypt and Yemen used poison gas in the 1960s. In 1988 Iraq's use of chemical weapons against the Kurds prompted other countries in the area, in particular Iran, Syria and Israel, to acquire such weapons.

Moscow's policy in this respect is a source of concern. After the break-up of the Soviet Union in 1991 Russia kept about 40,000 tonnes of chemical materials, accounting for two-thirds of the total worldwide. Through official sales or contraband it has become a key centre for their dissemination.

The 1972 Biological and Toxin Weapons Convention, which came into force in 1977, bans their development, production and storage, except for peaceful purposes. However in 2001

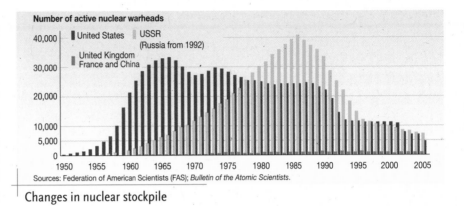

Changes in nuclear stockpile

the US opposed plans to introduce stricter controls for enforcement of the treaty. The 1993 Chemical Weapons Convention bans the development, production and storage of chemical weapons.

Dirty bombs, which combine conventional explosives and radioactive materials to contaminate the largest possible area, are the most likely vector for deliberate nuclear pollution by a terrorist group. These devices have not so far been used, so they do not count as WMDs, but they are nevertheless among the weapons terrorist groups might use.

On the web

> **Organisation for the Prohibition of Chemical Weapons (OPCW):**
www.opcw.org

> **International Atomic Energy Agency (IAEA):**
www.iaea.org

> **Carnegie Endowment for International Peace:**
www.carnegieendowment.org/npp/

> **United Nations Institute for Disarmament Research (UNIDIR):**
www.unidir.org

> **Center for Nonproliferation Studies (CNS):**
www.cns.miis.edu

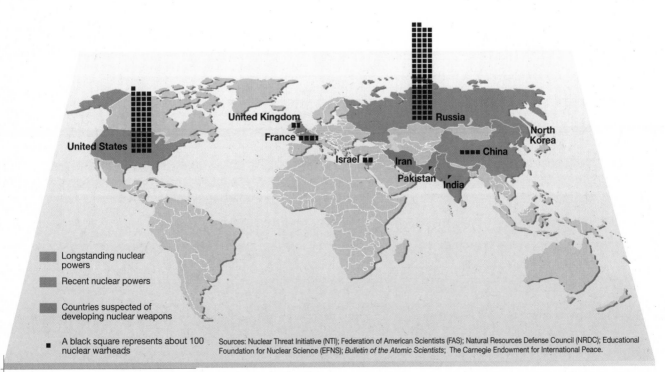

The real nuclear powers in 2005

Who causes industrial accidents?

The Johannesburg summit in 2002 emphasised the part that business would have to play in achieving sustainable development. But in many cases we are still waiting for tangible results, with large firms taking advantage of the laxity of national governments.

The tsunami that devastated South Asia had dramatic human consequences. At an environmental level it high-lighted the risks associated with nuclear facilities located on the coast. Seawater flooded the pump station of the 440 Megawatt power station at Kalpakkam in the state of Tamil Nadu, India, causing an emergency shut-down. On 9 August 2004, on the anniversary of the US attack on Nagasaki, an accident at the Mihama nuclear power station in Japan, 320 kilometres west of Tokyo, killed four people and injured seven others. Although it does not seem to have caused any radioactive contamination, it is yet another illustration of the safety problems posed by this industry and the lack of information available to the general public.

In recent years there has been much debate on nuclear safety in Japan. The press has reported that inspections of power stations have been hurried, with reports being forged. In April 2003 Tokyo Electric Power was ordered to shut down 17 of its reactors for safety reasons, following the discovery that it had concealed maintenance problems from the authorities.

In Russia very little information is available on the real state of repair of certain facilities. The situation in France is little better. It is for instance very difficult to obtain hard facts on the state of the nuclear power station at Fessenheim on the banks of the Rhine. Built in 1977 it is France's oldest nuclear power station, standing beside and the below the level of a canal (consequently subject to flooding) in an area with a relatively high seismic risk. At the beginning of 2004 seven incidents occurred there, contaminating 12 people.

The potential hazards of the chemical industry are no less disastrous. More than 20 years after the Bhopal disaster in India (on 3 December 1984), the site has still not been decontaminated. Worse still there is little indication firms in developed countries have learnt from the accident. Many western companies are still relocating operations to

Radioactive, chemical and biological hazards in Central Asia

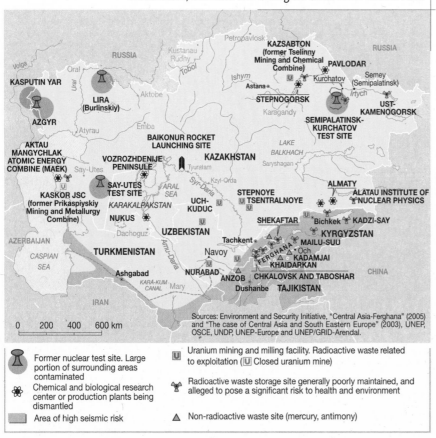

Sources: Environment and Security Initiative, "Central Asia-Ferghana" (2005) and "The case of Central Asia and South Eastern Europe" (2003), UNEP, OSCE, UNDP, UNEP-Europe and UNEP/GRID-Arendal.

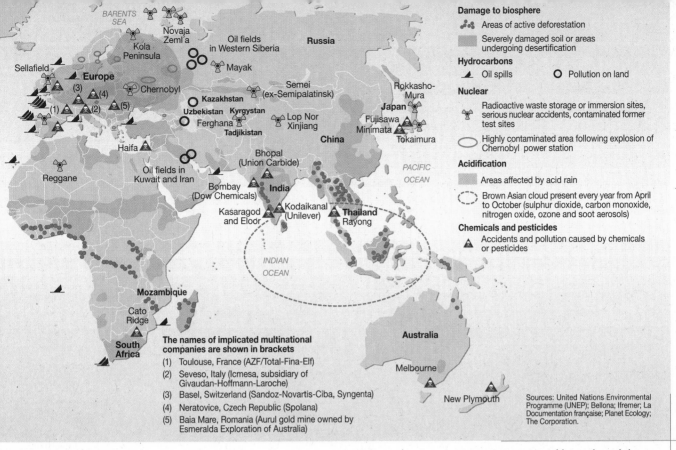

Damage to biosphere

🐾 Areas of active deforestation

▨ Severely damaged soil or areas undergoing desertification

Hydrocarbons

⬩ Oil spills   ○ Pollution on land

Nuclear

☢ Radioactive waste storage or immersion sites, serious nuclear accidents, contaminated former test sites

⬭ Highly contaminated area following explosion of Chernobyl power station

Acidification

▨ Areas affected by acid rain

◌ Brown Asian cloud present every year from April to October (sulphur dioxide, carbon monoxide, nitrogen oxide, ozone and soot aerosols)

Chemicals and pesticides

▲ Accidents and pollution caused by chemicals or pesticides

The names of implicated multinational companies are shown in brackets

(1) Toulouse, France (AZF/Total-Fina-Elf)

(2) Seveso, Italy (Icmesa, subsidiary of Givaudan-Hoffmann-Laroche)

(3) Basel, Switzerland (Sandoz-Novartis-Ciba, Syngenta)

(4) Neratovice, Czech Republic (Spolana)

(5) Baia Mare, Romania (Aurul gold mine owned by Esmeralda Exploration of Australia)

Sources: United Nations Environmental Programme (UNEP); Bellona; Ifremer; La Documentation française; Planet Ecology; The Corporation.

Major environmental hazards and damage

the South and upholding double standards, advocating environmental controls in the North and laissez-faire in the South. Polluters can avoid facing up to their responsibilities by concealing essential data, subcontracting dangerous jobs and relocating environmentally hazardous activities. One of the chemicals involved in the production process at Bhopal was phosgene, used to manufacture polyurethane foam. In France, though units using phosgene in Toulouse have closed (they were located close to the AZF plant destroyed by an explosion in 2001, but not directly affected) production continues in chemical plants in the Grenoble area. Yet there are several alternatives to phosgene available and plenty of scope for their development, enabling the product to be banned altogether.

The severe environmental damage caused by mining and oil-drilling are an additional source of concern. In French Guiana gold prospectors poison the air, ground and water of the Amazonian forest with mercury. Oil wells are a constant source of pollution, not only when accidents occur. They damage the deltas of rivers, such as the Niger in Nigeria or the Mahakam in Borneo. An estimated 600,000 tonnes of crude oil runs into the world's oceans every year, 30% coming from oil rigs, 60% from deliberate discharge and leakage from ships, and only 10% from oil spills.

In December 2004 a Malay tanker carrying 1.8m litres of crude oil broke up on the coast of Alaska. Fifteen years

■ On the web

> **Greenpeace:**
www.greenpeace.fr

> **International Campaign for Justice in Bhopal (ICJP):**
www.bhopal.net

> **International Rivers Network (IRN):**
www.irn.org

> **Natural Resources Defense Council (NRDC):**
www.nrdc.org

> **BioGems:**
www.savebiogems.org

> **Centre de documentation, de recherche et d'expérimentations sur les pollutions accidentelles des eaux (Cedre):**
www.le-cedre.fr

earlier the wreck of the Exxon Valdez spilt 40m litres of crude oil onto the same coastline. Since April 2005 European Union legislation has banned the use of single-hull tankers but the new rules raise the question of what is to be done with the old ships? Shipbreaking yards are some of the most polluted places in the world, perhaps the worst being Alang in India. Workers receive absolutely no guidance on the products (asbestos, heavy metals, oils) to which they are exposed. But France still intends to have the Clémenceau, one of its aircraft carriers, dismantled there. The Bush administration displays a similar disregard for the environment. Despite opposition in the US Senate it is still attempting to push through legislation authorising oil drilling in the Alaskan wilderness.

Efforts to achieve sustainable development

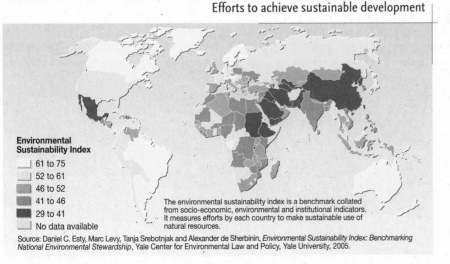

Environmental Sustainability Index

- ▨ 61 to 75
- ▨ 52 to 61
- ▨ 46 to 52
- ▨ 41 to 46
- ▨ 29 to 41
- ▨ No data available

The environmental sustainability index is a benchmark collated from socio-economic, environmental and institutional indicators. It measures efforts by each country to make sustainable use of natural resources.

Source: Daniel C. Esty, Marc Levy, Tanja Srebotnjak and Alexander de Sherbinin, *Environmental Sustainability Index: Benchmarking National Environmental Stewardship*, Yale Center for Environmental Law and Policy, Yale University, 2005.

Waste, recyclers and recycled

Promoting growth based on intensive productivity and consumption has major disadvantages, one of the most serious being the huge volume of waste produced and the problem of its disposal. Statistics fail to convey the full measure of the problem, particularly for industrial waste, now a commodity for international trade transported long distances around the world.

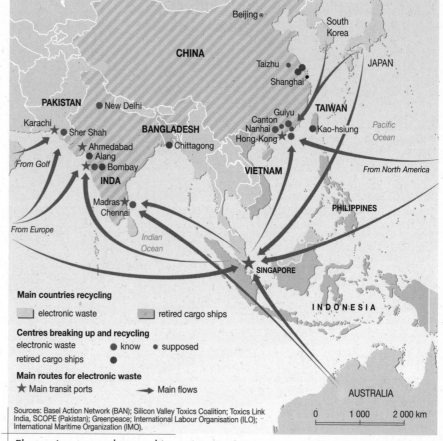

Electronic waste and cargo ships poisoning Asia

Sources: Basel Action Network (BAN); Silicon Valley Toxics Coalition; Toxics Link India, SCOPE (Pakistan); Greenpeace; International Labour Organisation (ILO); International Maritime Organization (IMO).

The mountain of waste resulting from mass production and consumption is growing increasingly cumbersome for our urban societies. Not only are their populations growing fast and consuming more, but the average service life of frequently "over-packaged" consumer goods is decreasing. Modern products contain an increasing number of materials that are difficult to break down, in particular certain plastics. As fewer resources are available for waste management than for the production of consumer goods, it will be an uphill struggle to slow the accumulation of waste, particularly in view of the rate at which some densely populated Asian countries are growing.

As for the import and export of waste, the first big surprise is how difficult it is to collate data. The Basel Convention, started in 1989 under the aegis of the United Nations, is an intergovernmental body tasked with monitoring and regulating the production and cross-border movement of waste. It provides figures that are difficult to interpret.

About 30 countries have so far refused to ratify the convention and do not publish any statistics. More surprisingly 110 of the 165 members states do not provide any data. Their number includes Norway, despite it boasting a highly progressive environmental policy. The lack of statistics is due to the complexity of the procedures for submitting figures and disparities between the assessment methods used by various countries.

However even incomplete data reveals several interesting points. The volume of waste in transit has substantially increased. For the 50 countries filing data the amount has increased from 2m tonnes in 1993 to 8.5m tonnes in 2001. Trade between countries belonging to the Organisation for Economic Cooperation and Development (OECD) accounted for three-quarters of the total. Almost all the waste was classified as "hazardous", a tricky term in itself, as even apparently harmless waste may generate risks if improperly managed.

During the 1980s environmental standards became much stricter in

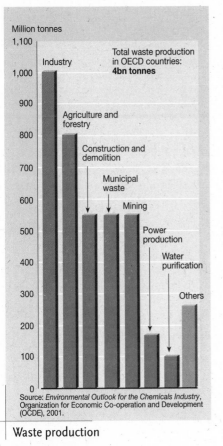

Million tonnes

Total waste production in OECD countries: **4bn tonnes**

Source: *Environmental Outlook for the Chemicals Industry*, Organization for Economic Co-operation and Development (OCDE), 2001.

Waste production

developed countries prompting an increase in waste traffic, particularly into Africa. Following a string of scandals (such as the odyssey in 1988 of the Italian cargo ship Zanoobia with its load of toxic waste) various international agreements were signed, regulating and in some cases banning the transport of waste to developing countries.

PROFIT MOTIVE

The flow of waste switched to countries in eastern Europe and the former Soviet Union (already struggling to cope with its own industrial past) then turned back to the main countries producing the waste. Two factors explain this change: first, the market for processing hazardous waste has considerable potential for specialist firms; second, it requires technology and infrastructure that are difficult to finance in poor countries. Rather than being seen as a problem hazardous waste now ranks as a source of potential profit.

Worse still rich countries send to Asia and Africa waste that causes too much pollution or generates too little profit, arguing that the material must be recycled anyway. The handling of

electronic waste (PCs, mobile phones, etc.) is typical of this trend. Volumes are rising steeply as service life shrinks. Several of the components contain toxic substances (cadmium, lead, mercury). Yet they are sent to China, India or South Africa to be dismantled and recycled. Not only does this process endanger the lives of workers, operating under conditions unsuited to the substances they are handling, but it also contaminates the atmosphere, ground and groundwater. Shipbreaking, an activity concentrated in China, India and Bangladesh, is yet another example of the same trend.

Many environmental experts condemn this type of recycling and are campaigning to promote alternative techniques. Production methods need to be rethought making allowance for the final disposal of goods. Waste must be processed locally to avoid its transport over long distances, and as much as possible must be recovered either for use as a raw material or an energy source. But above all we must rein in our consumption. This priority cuts across many environmental issues and represents the only credible solution for a planet which will be home to nine billion people by 2050.

The richer you are, the more you trash

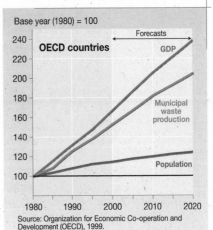

Source: Organization for Economic Co-operation and Development (OECD), 1999.

■ On the web

› Secretariat of the Basel Convention on transboundary movements of hazardous waste: www.basel.int

› Basel Action Network (BAN): www.ban.org

› «Exporting Harm : The High-Tech Trashing in Asia» (Basel Action Network/ Silicon Valley Toxics Coaliton): www.crra.com/ewaste/ttrash2/ttrash2/#ewaste

› Electronic Waste Guide: www.ewaste.ch

› «Is There A Decent Way to Break Up Ships ?» (International Labour Organisation – ILO): www.ilo.org/public/english/dialogue/sector/papers/shpbreak/

Largely disregarded agreements

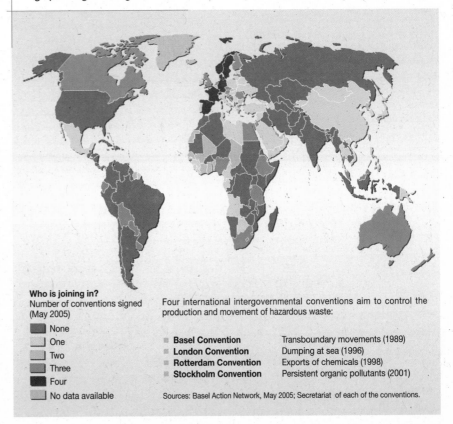

Who is joining in?
Number of conventions signed (May 2005)

- None
- One
- Two
- Three
- Four
- No data available

Four international intergovernmental conventions aim to control the production and movement of hazardous waste:

Basel Convention	Transboundary movements (1989)
London Convention	Dumping at sea (1996)
Rotterdam Convention	Exports of chemicals (1998)
Stockholm Convention	Persistent organic pollutants (2001)

Sources: Basel Action Network, May 2005; Secretariat of each of the conventions.

The South depends on its exports

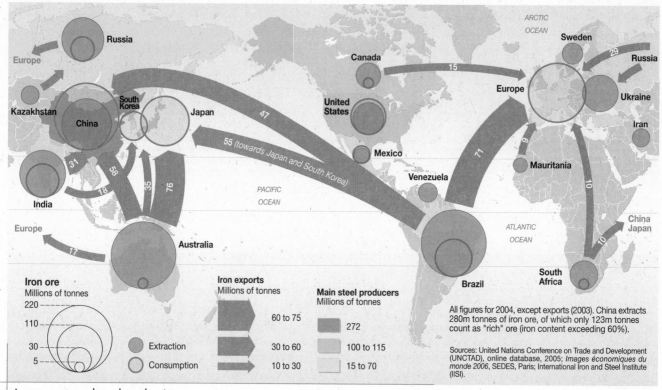

Iron ore
Millions of tonnes
220
110
30
5

○ Extraction
○ Consumption

Iron exports
Millions of tonnes
60 to 75
30 to 60
10 to 30

Main steel producers
Millions of tonnes
272
100 to 115
15 to 70

All figures for 2004, except exports (2003). China extracts 280m tonnes of iron ore, of which only 123m tonnes count as "rich" ore (iron content exceeding 60%).

Sources: United Nations Conference on Trade and Development (UNCTAD), online database, 2005; *Images économiques du monde 2006*, SEDES, Paris; International Iron and Steel Institute (IISI).

Iron exports and steel production

Since the 1970s the price of raw materials has followed a downward trend, subject to great instability. But developing countries, heavily in debt and dependent on their exports, are increasingly reluctant to bow to the demands of rich countries, as the failure of the Cancun negotiations demonstrated.

According to the United Nations Conference on Trade and Development the overall distribution of exports from developing countries has changed a great deal in the last 20 years. About 70% of such exports are now manufactured goods – particularly from Asian countries – whereas basic commodities used to account for three-quarters of the total. However these figures conceal major disparities between different parts of the world. Africa has scarcely benefited from the boom in exports of manufactured goods, which still only represent an average of 30% of the total, compared with 20% in 1980.

The price of basic commodities, which started rising in the early 1960s, began to fall from 1974 onwards, in a series of sudden drops and brief rallies. With the financial crisis in Asia the period from 1997 to 2001 saw an

overall drop in prices, losing almost 53% of their value. Basic commodities were suddenly worth half as much as manufactured goods.

The main reason for the drop in prices is market saturation. With the huge increase in their deficits in 1960-70 developing countries had to boost exports to earn the hard currency needed to repay loans. In many cases they specialised in two or three commodities, becoming very dependent on them and competing with other countries in a similar predicament, which in turn pushed prices down. This combination of factors played a key role in the debt crisis, which enabled multinational companies and owners of capital assets to take control of the world economy. Over the last 25 years the North has imposed structural adjustment programmes, with the removal of price controls, on the most

Raw materials price index
Base year (1985) = 100

Source: United Nations Conference on Trade and Development (UNCTAD), online database, 2005.

Falling prices

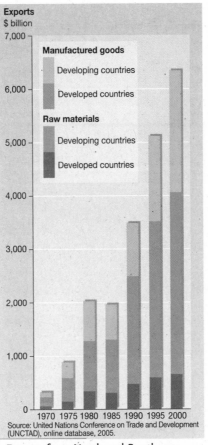

Exports
$ billion

Manufactured goods
Developing countries
Developed countries

Raw materials
Developing countries
Developed countries

Source: United Nations Conference on Trade and Development (UNCTAD), online database, 2005.

Exports from North and South

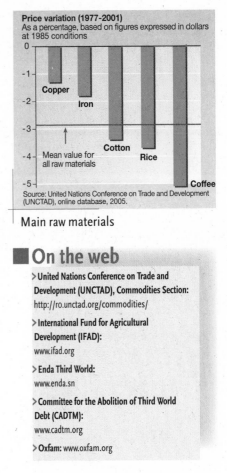

Price variation (1977-2001)
As a percentage, based on figures expressed in dollars at 1985 conditions

Copper
Iron
Cotton
Rice
Coffee

Mean value for all raw materials

Source: United Nations Conference on Trade and Development (UNCTAD), online database, 2005.

Main raw materials

On the web

> United Nations Conference on Trade and Development (UNCTAD), Commodities Section:
http://ro.unctad.org/commodities/

> International Fund for Agricultural Development (IFAD):
www.ifad.org

> Enda Third World:
www.enda.sn

> Committee for the Abolition of Third World Debt (CADTM):
www.cadtm.org

> Oxfam: www.oxfam.org

heavily indebted countries. This has further increased the dependence of poor countries on basic commodities, making them all the more vulnerable to fluctuations in the world market.

UNFAIR SUBSIDIES

In the case of agricultural products, variations in prices are due to natural and weather conditions, and to political instability (the price of cocoa rose, for instance, at the end of 2002 following unrest in the Ivory Coast) or the arrival of new producers (such as Vietnam for coffee). But the unfair subsidies paid by the United States and the European Union to their farmers (for products such as cotton, sugar or meat) led to deadlock at the World Trade Organisation (WTO) conference in Cancun in September 2003. The US is the world's largest cotton exporter thanks to its massive subsidies ($3.9bn in 2001-2). Yet, according to the International Cotton Advisory Committee, it costs $0.21 to produce a pound of cotton in Burkina Faso, compared with $0.73 in the US. The impact on human development is immediate. In Bénin, for example, the falling price of cotton (down 35% in 2001) led to a 4% increase in poverty.

Furthermore rich countries levy almost no customs duty on raw materials, discouraging poor countries from diversifying their economy to produce manufactured goods.

The sudden upturn in the price of key raw materials since 2004 is due to considerable growth in Chinese demand (particularly for oil), political instability in Iraq (following the US invasion) and Yukos' legal problems in Russia.

The rich countries still control finance and transport, and wield considerable influence over the WTO, but future north-south negotiations will hinge on the issue of agricultural raw materials. Cancun saw the emergence of groups of developing countries (in particular the Group of 22), adding to the difficulties of the US and Europe. But a drop in rich countries' export subsidies would not necessarily help the poorest developing countries, particularly in Africa. The top priority for them is to defend their food sovereignty, in other words their right to define agricultural and land policies that are socially and economically appropriate to their unique circumstances.

The biggest exporters

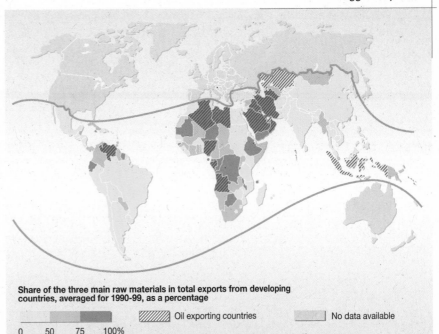

Share of the three main raw materials in total exports from developing countries, averaged for 1990-99, as a percentage

0   50   75   100%

Oil exporting countries   No data available

Source: United Nations Conference on Trade and Development (UNCTAD), online database, 2005.

losing the battle against

In 2000 there were 852 million undernourished people on Earth. Over the last five years their number has increased every year by about 4 million. Without a radical change of course we will not achieve the United Nations Millennium Development Goal (of reducing by half the proportion of people suffering from hunger by 2015). The reasons for this failure are all too familiar.

More than 20 million children with low birth weight are born every year in developing countries. The subsequent growth of one child in three is hindered by chronic malnutrition. The damage inflicted is considered irreversible. According to the United Nations' Food and Agriculture Organisation "the number of food emergencies has been rising over the past two decades, from an average of 15 per year during the 1980s to more than 30 per year since the turn of the millennium. Most of this increase has taken place in Africa, where the average number of food emergencies each year has almost tripled."

Drought is the main natural cause. Ready access to water increases yields and makes it easier for people to secure a proper food supply. Irrigated farm land, which represents 17% of the total area under cultivation, produces 40% of all food. Other factors, such

as flooding, frost or locusts also come into play. But human factors (conflicts, movement of population, economic decisions) are increasingly involved, causing more than 35% of food emergencies in 2004, compared with only 15% in 1992. As the FAO explains: "In many cases, natural and human-induced factors reinforce each other. Such complex crises tend to be the most severe and prolonged. Between 1986 and 2004, 18 countries were 'in crisis' more than half of the time. War or economic and social disruptions caused or compounded the crises in all 18."

In economic terms the free market policies imposed by the International Monetary Fund and the World Bank, with the consent of local leaders, are responsible for a large part of the increase in food insecurity. In particular they demanded an end to subsidies on essential foodstuffs. As a

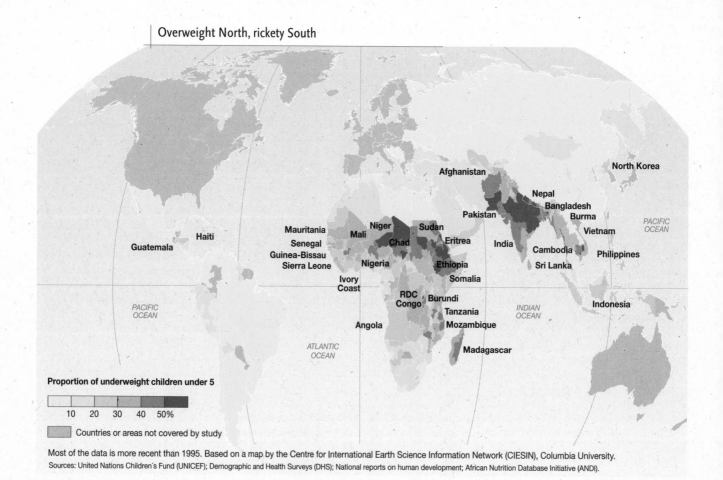

Overweight North, rickety South

Proportion of underweight children under 5

10 20 30 40 50%

Countries or areas not covered by study

Most of the data is more recent than 1995. Based on a map by the Centre for International Earth Science Information Network (CIESIN), Columbia University.
Sources: United Nations Children's Fund (UNICEF); Demographic and Health Surveys (DHS); National reports on human development; African Nutrition Database Initiative (ANDI).

hunger

priority all available income must be directed to repaying foreign debt. The brutal deregulation of the economy in developing countries, which became a form of dogma under the structural adjustment plans and is still keenly defended by the World Trade Organisation, disrupted farming production in the South. Export subsidies in rich countries and unfair rules of world trade made the situation even worse. Furthermore foreign aid for agriculture has substantially decreased in real terms since 1980. In Africa, for instance, foreign aid per farming worker is only a quarter of the amount it was in 1982. Above all donors allocate aid in line with global criteria, so the countries most in need rarely benefit.

DECLINING PRODUCTION

The spread of HIV-Aids is also proving a decisive factor. In southern Africa at least one in five people working in agriculture will die before 2020, severely jeopardising the population's food supply. Growth in global crop and livestock production has been slowing in recent years and according to the FAO, "the slow rate of growth in 2002 of less than 1% at the global level implies a reduction in output in per capita terms". Sub-Saharan Africa is in a critical predicament, as "the only region that has not seen increases in

per capita food production over the last three decades; after a pronounced decline in the course of the 1970s and early 1980s, per capita food production has stagnated and is still at levels reported two decades ago."

The most extreme instance is the Democratic Republic of Congo. Despite having rich natural resources, 71% of its inhabitants are suffering from hunger. Thirty-five countries (of which 24 in Africa) are facing severe food shortages, a clear demonstration of the shortcomings of the present system.

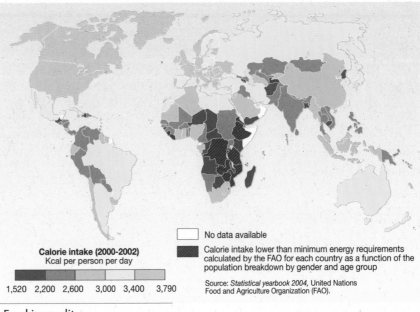

Calorie intake (2000-2002)
Kcal per person per day

1,520 2,200 2,600 3,000 3,400 3,790

☐ No data available

▨ Calorie intake lower than minimum energy requirements calculated by the FAO for each country as a function of the population breakdown by gender and age group

Source: *Statistical yearbook 2004*, United Nations Food and Agriculture Organization (FAO).

Food inequality

■ On the web

> United Nations Food and Agriculture Organisation (FAO):
www.fao.org/es/

> World Food Programme (WFP):
www.wfp.org

> World Resources Institute (WRI):
www.wri.org

> Action Against Hunger:
www.actioncontrelafaim.org

> Comité catholique contre la faim et pour le développement (CCFD) :
www.ccfd.asso.fr

Inadequate production

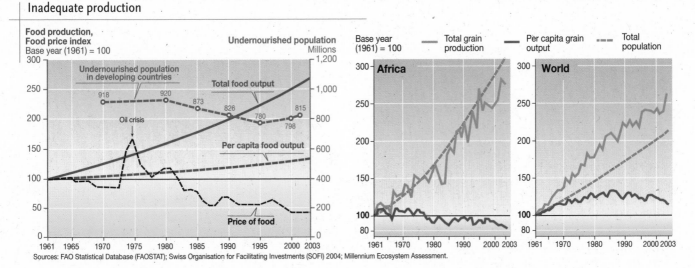

Food production, Food price index
Base year (1961) = 100

Undernourished population in developing countries
918 920 873 826 780 798 815
Oil crisis
Total food output
Per capita food output
Price of food

Undernourished population Millions

Base year (1961) = 100 — Total grain production — Per capita grain output --- Total population

Africa

World

Sources: FAO Statistical Database (FAOSTAT); Swiss Organisation for Facilitating Investments (SOFI) 2004; Millennium Ecosystem Assessment.

GM organisms, too much, too soon

The issue of genetically modified organisms draws together strands from the debate on the global market and the concept of progress. It is a perfect illustration of how market forces come into play much more quickly than the precautions that seem appropriate given the current state of research. We are consequently already eating genetically engineered foodstuffs without it being possible to guarantee they are entirely safe.

Growing numbers of consumers are eating increasing amounts of GM produce. Environmental activists may have made much publicised attempts to halt trials of GM crops (in particular in France) but by the end of 2004 some 8.2 million farmers worldwide were growing GM crops They cover 81m hectares, up by 20% on 2003-4, and already occupy 5.4% of the world's cultivated land. Long restricted to developed countries (North America) they are gaining ground in poor countries, particularly in India and China where substantial resources have been earmarked for their development.

Though production has focussed mainly on soy beans, maize, cotton and rape seed, GM foodstuffs will sooner or later find their way onto our plates. In view of the likelihood of natural or accidental contamination and the complexity of agrifood supply chains in an increasingly global market, it is foolhardy to imagine that any part of the process can be completely sealed off from the rest. Apart from products

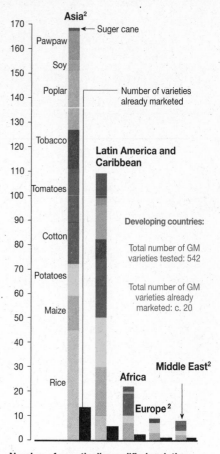

Number of genetically modified varieties being experemented in developing countries[1]

1. Ten main species tested in laboratories or open fields
2. Developing countries only

Source: FAO-BioDeC, 2005.

South acting as guinea pig

manufactured at considerable cost and guaranteed GM-free, we all eat food containing genetically modified ingredients. Europe acknowledged this state of affairs when it authorised their presence providing it was mentioned on the label (for quantities exceeding 0.9% of a product). This supposedly "protectionist" measure prompted an outcry in the United States. Either way, it is up to consumers to shoulder their responsibilities.

The first generation of GM crops are of little direct benefit to consumers,

Growing output

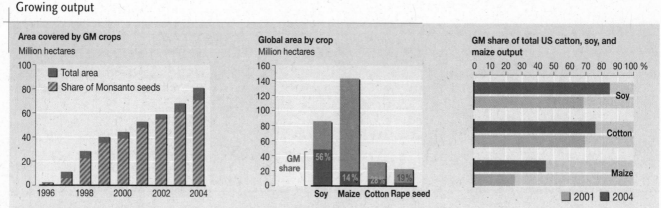

Sources: International Service for the Acquisition of Agri-Biotech Applications (ISAAA), 2004; Monsanto, 2005; Clive James; United States Department of Agriculture (USDA).

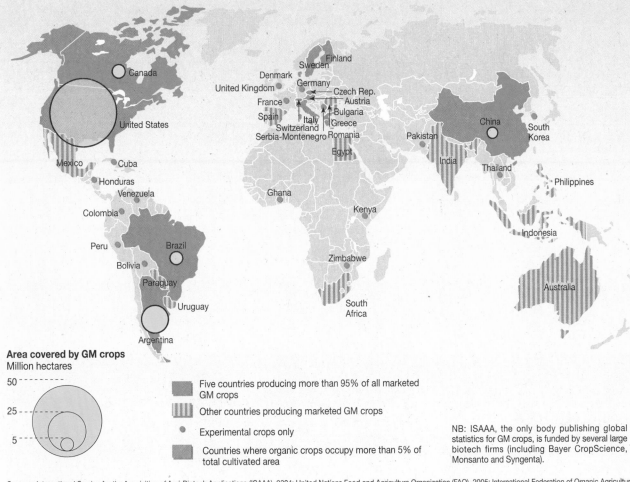

Area covered by GM crops
Million hectares

50
25
5

Five countries producing more than 95% of all marketed GM crops

Other countries producing marketed GM crops

Experimental crops only

Countries where organic crops occupy more than 5% of total cultivated area

NB: ISAAA, the only body publishing global statistics for GM crops, is funded by several large biotech firms (including Bayer CropScience, Monsanto and Syngenta).

Sources: International Service for the Acquisition of Agri-Biotech Applications (ISAAA), 2004; United Nations Food and Agriculture Organization (FAO), 2005; International Federation of Organic Agriculture Movements (IFOAM), 2005; Institute for Health and Consumer Protection (IHCP) of the European Commission's Joint Research Centre (JRC).

The five largest GM producers

though this may change in the future with specially engineered plants capable of trapping pollution or surviving droughts. But above all it is an uphill struggle forming an opinion amidst the conflicting views offered by experts, regardless of whether debate focuses on the consequences of GM crops for the environment, public health or the economy. As with so many issues related to living organisms, it is hard to distinguish between rational and emotional responses.

Ecologists, seed merchants and even scientists are in complete disagreement about the environmental impact of GM organisms. Their dissemination by pollen is a potential risk for biodiversity. Observations in the United Kingdom and Germany have confirmed the risk of a "bio-invasion". In Mexico, where humans first cultivated maize, the discovery that foreign genes introduced by US imports had contaminated traditional strains prompted a public outcry. Rather than attempting to explain what is going on, GM advocates maintain that such crops reduce the need for fertilisers

and pesticides, limit soil erosion and enable simpler farming techniques.

There is still no certainty about the long-term impact of GM organisms on public health, no systematic studies having been carried out, even in the US where consumers have been eating GM foods for years. Some experiments – open to doubt – suggest that changes have been detected in the blood and kidneys of laboratory rats. On the other hand some people claim that GM crops reduce mycotoxin (a form of fungus) contamination, which in turn lowers the risk of cancer.

Even the economic benefits are open to dispute. In South Africa the spread of insect pests (tarnished plant bugs) wiped out any benefit some small producers might have derived from investing expensive GM seeds. In other cases (fruit rot, vine growth malformation) a clear improvement in yields has been observed. Either way, GM crops seem certain to increase poor countries' dependence on a few giant firms such as Monsanto or Bayer, long before they end the uneven distribution of food across the planet.

The precautions taken in Europe – special labelling, refusal by about 50 regions to authorise GM crops – and the destruction of trial crops by French environmental activists, are holding back research into their impact on biodiversity without stopping imports of genetically modified seeds or products. There is a genuine risk that GM organisms become so widespread that they pass the point of no return, before governments have taken even the most elementary precautions, equivalent to the tests preceding the public launch of new drugs.

■ On the web

> **Inf'OGM:** www.infogm.org

> **GeneWatch:** www.genewatch.org

> **Organic Consumers:**
www.organicconsumers.org

> **Comité de recherche et d'information indépendantes sur le génie génétique (CRRI-GEN):**
www.crii-gen.org

> **French government site on GM crops:**
www.ogm.gouv.fr

Urban culture, with its values, fads and fashions, now governs the entire planet. But from one continent to the next the form and pace of urban development vary, even if motor vehicles and their attendant infrastructure are omnipresent.

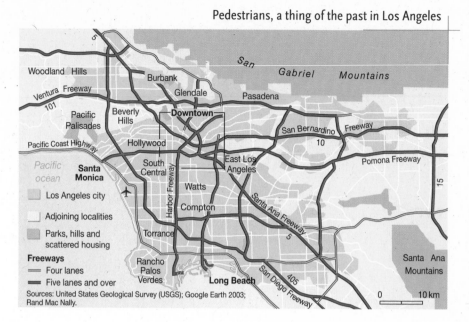

Pedestrians, a thing of the past in Los Angeles

Sources: United States Geological Survey (USGS); Google Earth 2003; Rand Mac Nally.

Urban development trends

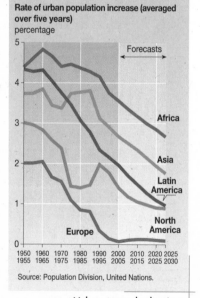

Rate of urban population increase (averaged over five years)
percentage

Source: Population Division, United Nations.

Urban growth slowing

At the beginning of the 20th century there were 11 cities with a population exceeding one million people. Their number, which reached 80 by 1950, 276 in 1990 and almost 400 in 2000, now seems likely to hit 550 by 2015. Urban development does not only mean the concentration of statistically quantifiable population groups. It also involves profound changes, subjecting expanses of land, the people who live there and their institutions, to urban culture. It "urbanises" social mores and generalises a particular lifestyle, a society of individuals, whose mobility reflects their relative autonomy, based on a uniform, repetitive timetable.

According to the historian Fernand Braudel "towns are a fortunate accident of history" which coincided with the birth of agriculture, some 8,000 to 10,000 years ago. Now at the start of the 21st century humanity faces a new situation, with the foreseeable decline of farming communities and the disappearance of rural culture.

Urban development has not affected all continents evenly. Most of Europe's population lives in an urban sprawl, with towns and their immediate vicinity forming dynamic networks. Only Greater London, Moscow and the Paris area are home to several million people. In contrast America (North and South) has many vast conurbations with millions of people (Mexico City, Sao Paulo, Buenos Aires, New York and Los Angeles each have populations exceeding 15 million). The pace of urban development in Asia is even more rapid and by 2020 it will have a dozen or so giant metropoles (such as Mumbai, Karachi, Shanghai, Dacca, Jakarta or Tokyo) each with nearly 20 million inhabitants. Three quarters of Australia-Oceania is already urbanised. As for Africa the same process is at work, but operating on various scales. Several huge urban areas have nevertheless formed, such as Lagos (300,000 inhabitants in 1950, almost 10 million now), Kinshasa or Cairo.

Cities are places of stark contrasts and there may be sudden changes in social standing between one district and the next. In many cases most of the population lives in slums and shanty towns – referred to using various terms

Urban growth from 1990 to 2003

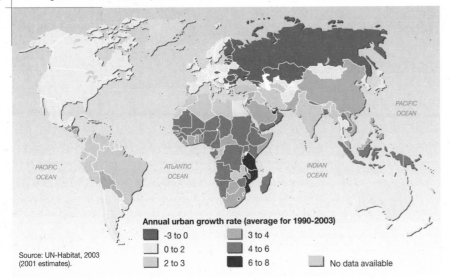

Annual urban growth rate (average for 1990-2003)
- -3 to 0
- 0 to 2
- 2 to 3
- 3 to 4
- 4 to 6
- 6 to 8
- No data available

Source: UN-Habitat, 2003
(2001 estimates).

Geography of shanty towns

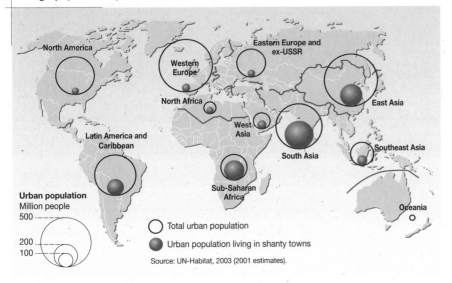

Urban population
Million people
- 500
- 200
- 100

- Total urban population
- Urban population living in shanty towns

Source: UN-Habitat, 2003 (2001 estimates).

■ On the web

> **World Urbanization Prospects (United Nations):**
http://esa.un.org/unup/

> **Population data:**
www.populationdata.net/villes.html

> **DEWA-GRID-Europe, cities from space:** www.grid.unep.ch/activities/global_change/cities_from_space.fr.php

> **Centre population et développement (CEPED):**
http://ceped.cirad.fr

representatives of all the top law firms, advertising agencies and accountants, and high-powered communications and transport facilities (intermodal hubs). It is here that the global economy is controlled.

Setting aside growing social strife, environmental problems are becoming increasingly pressing. Neighbouring districts are obliged to compete for dwindling water resources (one in four of the world's inhabitants must cope with chronic water shortages, while industrial farming squanders this precious resource and households in rich countries indulge in gross over-consumption). Similarly the energy consumption of western cities is constantly increasing – heating is no longer enough, we need air-conditioning too – placing growing demands on non-renewable energy sources. Ever more widespread use of motor vehicles is aggravating atmospheric pollution. Noise caused by machinery and over-population makes silence and solitude unthinkable. Finally city-dwellers are gradually being deprived of access to public spaces and amenities, essential to an enduring sense of belonging to a larger social whole. Is what started as a "fortunate accident" in the process of turning into a tragic mistake?

including *favelas, colonias proletarias, kampong* and *gecekondu.* By giving the occupants of shanty towns a legal claim to the plot of land they occupy it is possible to improve housing conditions and restrict the power of organised crime. At the other end of the social scale, residential enclaves are increasingly turning into gated communities, catering for the demands of the upper middle classes of Los Angeles, Rio, Istanbul and New Delhi, not to mention Moscow, Rome and Toulouse. Sprawling suburbs are another dominant feature of much urban development, with built-up areas, of various density, extending over large distances without any real sense of unity or identity.

Global cities boast stock markets and the headquarters of large firms,

Shift to the cities

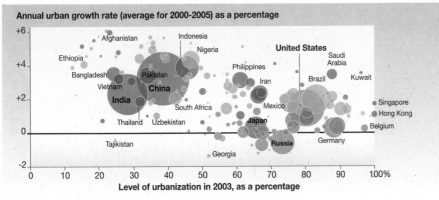

Annual urban growth rate (average for 2000-2005) as a percentage

Level of urbanization in 2003, as a percentage

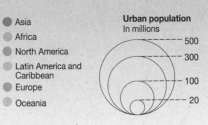

- Asia
- Africa
- North America
- Latin America and Caribbean
- Europe
- Oceania

Urban population
In millions
- 500
- 300
- 100
- 20

Source: United Nations Department of Economic and Social Affairs, Population Division, 2005.

Planet in peril

Widening health care

Unequal access to health care is the cruellest, most widespread attack on human integrity. Coming on top of longstanding differences in standards of living, the balance of power that lets North dominate South inflicts chronic bad health on whole countries, sapping any attempt at development.

The disparity in the wealth of various parts of the world explains to a large extent the differences in their overall state of health. A little Japanese girl born in 2005 has an average life expectancy of 85 years, more than twice the life span of a baby girl in Zimbabwe (36 years, 2003 data). The reasons for this scandalous imbalance are well known: poverty, inadequate medical facilities, failure to control epidemics, and the high financial return expected of investment in medical research. The quality of treatment for the commonest complaints (measles, asthma, heart disease, psychiatric difficulties, cancer) is simply lower in poor countries, and as a result they kill or disable many more people.

Every day HIV-Aids kills 8,000 people (mainly young adults) and malaria another 3,000 (mainly children). Tuberculosis claims 6,000 more lives. These three big pandemics cause six million deaths every year, mainly in the poorest communities, in particular in sub-Saharan Africa. But the area

they affect is spreading.

The United Nations Security Council and the United States government (National Security Council) have stressed that the health crisis is threatening the political stability of many countries and might damage US interests. Yet the world has all the resources it needs to solve the problem. Rather than spending $240bn on the war in Iraq (the final cost will certainly exceed this figure) the US could have gone a long way to achieving the UN Millennium Goal for health care.

But, disregarding for a moment our lack of humanitarian ambition and strategic vision, even more insidious processes are at work sapping the medical resources of countries that try against the odds to develop effective ways of combating ill health.

First, the leading drug firms, aka Big Pharma, make the whole world pay for their increasingly financially oriented business model. They maintain that only scrupulous compliance with patent rules can secure present

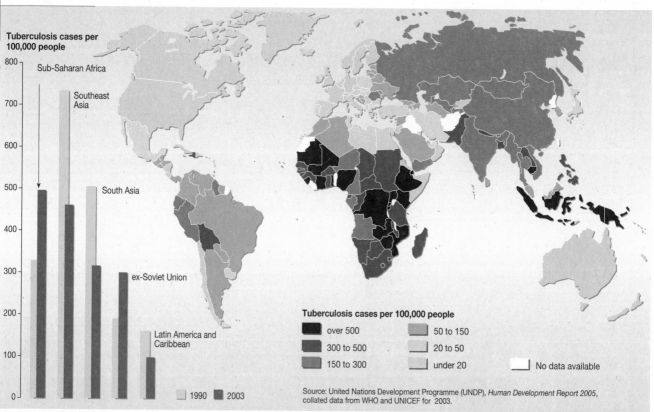

Increasing incidence of tuberculosis in sub-Saharan Africa and CIS countries

Tuberculosis cases per 100,000 people

Sub-Saharan Africa • Southeast Asia • South Asia • ex-Soviet Union • Latin America and Caribbean

☐ 1990   ■ 2003

Tuberculosis cases per 100,000 people

- over 500
- 300 to 500
- 150 to 300
- 50 to 150
- 20 to 50
- under 20
- No data available

Source: United Nations Development Programme (UNDP), *Human Development Report 2005*, collated data from WHO and UNICEF for 2003.

gap

and future investment in medical research. But steeply rising expenditure in this field above all focuses on more "profitable" diseases. Through active lobbying the drug firms have convinced the US administration, supported to a large extent by the European Union, to exert considerable pressure on countries, such as India, Brazil and South Africa, to discourage them from using generic drugs. Supplies of free medicine must comply with market rules. In its determination to prevent India from producing generic drugs, the North literally bought it off with trade concessions in other fields.

Second, political and religious considerations may condition the allocation of international aid. The quite substantial US contribution to combating Aids is linked to President Bush's pro-life policies.

Last but not least, with public funds in increasingly short supply due to the structural adjustment policies imposed by international donors, it is difficult to retain the health workers needed to provide a proper service. The governments of poor, debt-ridden countries are under pressure to cut welfare spending and limit public-sector pay packets, whereas rich countries are busy attracting staff trained elsewhere at no cost to them. More than 23% of

doctors working in the US trained in foreign countries where, in the vast majority of cases (86%) salaries are much lower.

If the average life expectancy in Zimbabwe is only 36 years it is also because three quarters of the doctors trained there emigrate at the end of their studies, fleeing Aids, pitiful pay and political persecution. This brain

drain is equivalent to poor countries paying almost $500m in aid to rich countries every year. It also increases inequality at home, between people living in the country (with no chance of finding a doctor) and city dwellers, between the poor obliged to make do with a totally inadequate public service and the rich who can afford private treatment.

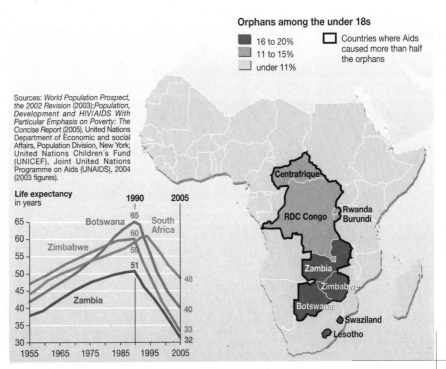

Orphans among the under 18s

- 16 to 20%
- 11 to 15%
- under 11%
- Countries where Aids caused more than half the orphans

Sources: *World Population Prospect, the 2002 Revision* (2003);*Population, Development and HIV/AIDS With Particular Emphasis on Poverty: The Concise Report* (2005), United Nations Department of Economic and social Affairs, Population Division, New York; United Nations Children´s Fund (UNICEF), Joint United Nations Programme on Aids (UNAIDS), 2004 (2003 figures).

Southern Africa decimated by HIV-Aids

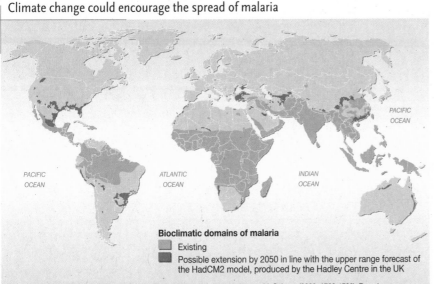

Climate change could encourage the spread of malaria

Bioclimatic domains of malaria
- Existing
- Possible extension by 2050 in line with the upper range forecast of the HadCM2 model, produced by the Hadley Centre in the UK

Source: Rogers and Randolph, The global spread of malaria in a future, warmer world, *Science* (2000: 1763-1766). Based on a map by Hugo Ahlenius, UNEP/GRID-Arendal, Norway.

■ On the web

> **Treatment Action Campaign (TAC):**
www.tac.org.za

> **Doctors Without Borders (MSF):**
www.msf.fr

> **Medact, Global Health Watch:**
www.medact.org/hpd_global_health_watch.php

> **Consumer Project on Technology (CPTech):**
www.cptech.org

> **Network for medicines and development (ReMeD):**
www.remed.org

> **World Health Organisation (WHO):** www.who.org

China a key factor in tomorrow's climate

With almost a quarter of the world's population China is putting massive pressure on its natural resources, particularly in the eastern part of the country, home to two-thirds of the people. In just a few decades its geography has changed.

There is no longer any doubt that a major ecological change is underway. In 1997 the Yellow river (Huang He in Chinese), one of the world's largest watercourses and the birthplace of civilisation in northern China, dried up before it reached the sea, simply because too much water had been drawn off to feed irrigation systems in upstream provinces. In 2003 it recorded its lowest level in 50 years. Sixty of China's 560 rivers have already dried up or are nearing depletion. Meanwhile work is underway on a gigantic project to build a 1,600 kilometre long, north-south canal connecting the Yangtze river to the Yellow river, in an attempt to remedy the water shortage in Beijing and the northern provinces.

A GLUTTON FOR ENERGY

Erosion now affects half of northern China. Persistent drought whips up increasingly frequent and destructive sand storms. In the east the desert is advancing at the rate of 2,500 kilometres a year. The first dunes are only 70 kilometres from Beijing.

The rise in the average temperature (1.5°C since the 1950s) is melting the Himalayan glaciers which feed the Yellow river. The Halong glacier, for instance, has shrunk by almost a fifth in 30 years. Downstream soil erosion is reducing crop yields and, to make matters worse, Chinese farmers must also cope with falling groundwater levels.

The drought in the north goes hand in hand with floods in the south. Flooding by the Yangtze in 1998 – aggravated

China is fast becoming the workshop of the 21st century world. But a shortage of raw materials abroad and increasingly serious environmental problems at home are already threatening continued growth.

by logging of 85% of all the trees along the river bank – claimed 4,000 lives and left 18 million homeless. Since the disaster the authorities have banned further logging, a measure that has slowed deforestation in the mountains but forced industry to import large quantities of wood to cover its growing needs. The flooding in China has thus had a catastrophic impact on forests in Russia and Southeast Asia, the source of much of the timber imported illegally into the country.

China is the world's top consumer of fertilisers. It is also the fifth largest producer of genetically modified crops, with GM cotton and soy fields covering 3.7m hectares. Despite the use of increasingly industrial farming techniques food production is failing to keep up with demand. If in some hypothetical future Chinese consumption reached a level equivalent to current American figures, the country would absorb the equivalent of two-thirds of the wheat and three-quarters of the meat produced in the world in 2004.

China is a glutton for energy too. It boasts the world's second largest hydro-electric power facilities. The construction of the largest ever dam – the Three Gorges scheme on the Yangtze river – involves moving 1.2

million people to new towns. The future reservoir will cover 58,000 square kilometres, an area larger than Switzerland. The dam is located at the heart of one of the regions of China with the greatest biodiversity and will affect thousands of species, including the unique Baiji dolphin.

With the manufacture of 2.5m private cars in 2004, China is fast becoming the world's top source of greenhouse gases, while pushing up the price of oil and raw materials. It consumes 40% of the world's coal output, 25% of all steel and nickel, 19% of aluminium. It has supplanted Japan as the world's second largest oil consumer, behind the United States. With 5,000 kilometres of new roads every year, it leads the world for infrastructure development, the aim being to have a network of motorways stretching over 70,000 kilometres by 2010.

In 2004 the Chinese Academy of Social Sciences published a report which suggested that if allowance was made for the true social and environmental cost of development at this pace it would knock two points off the growth rate. The key challenge for China is how it will direct future expansion, a decision that will affect the rest of the planet.

■ On the web

> **State Environmental Protection Agency:**
www.zhb.gov.cn/english

> **National Coordination Committee on Climate Change:**
www.ccchina.gov.cn

> **China Environment Daily:**
www.cenews.com.cn

> **Worldwatch Institute:**
www.worldwatch.org/features/chinawatch

> **Greenpeace Yellow River Source:**
www.yellowriversource.org

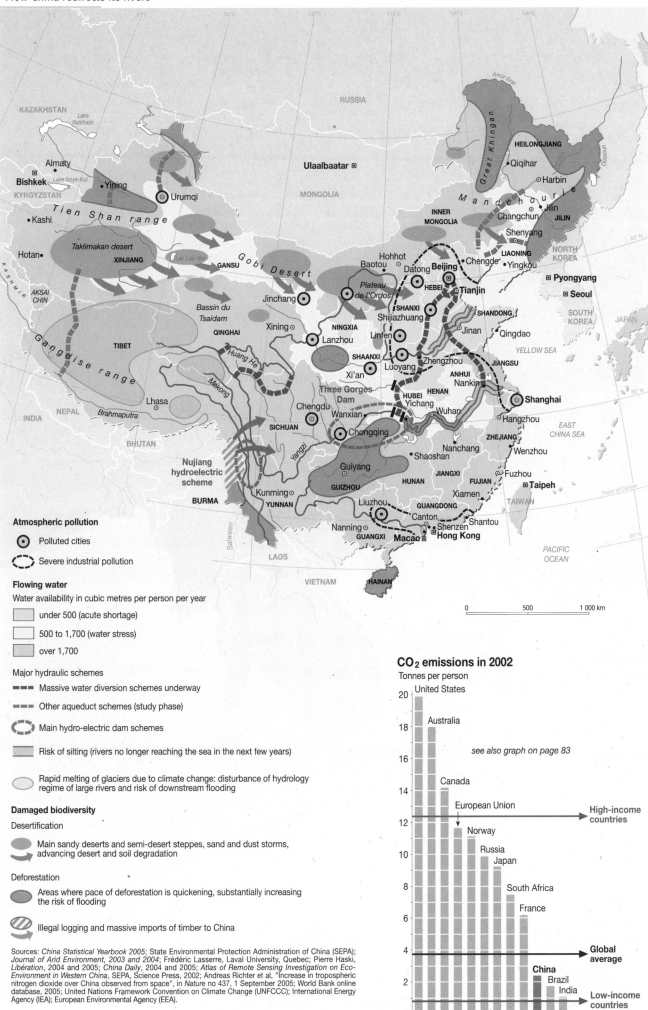

Atmospheric pollution

◉ Polluted cities

◌ Severe industrial pollution

Flowing water

Water availability in cubic metres per person per year

under 500 (acute shortage)

500 to 1,700 (water stress)

over 1,700

Major hydraulic schemes

▬ ▬ ▬ Massive water diversion schemes underway

▬ ▬ ▬ Other aqueduct schemes (study phase)

◌ Main hydro-electric dam schemes

Risk of silting (rivers no longer reaching the sea in the next few years)

Rapid melting of glaciers due to climate change: disturbance of hydrology regime of large rivers and risk of downstream flooding

Damaged biodiversity

Desertification

Main sandy deserts and semi-desert steppes, sand and dust storms, advancing desert and soil degradation

Deforestation

Areas where pace of deforestation is quickening, substantially increasing the risk of flooding

Illegal logging and massive imports of timber to China

Sources: *China Statistical Yearbook 2005*; State Environmental Protection Administration of China (SEPA); *Journal of Arid Environment, 2003 and 2004*; Frédéric Lasserre, Laval University, Quebec; Pierre Haski, *Libération*, 2004 and 2005; *China Daily*, 2004 and 2005; *Atlas of Remote Sensing Investigation on Eco-Environment in Western China*, SEPA, Science Press, 2002; Andreas Richter et al, "Increase in tropospheric nitrogen dioxide over China observed from space", in *Nature* no 437, 1 September 2005; World Bank online database, 2005; United Nations Framework Convention on Climate Change (UNFCCC); International Energy Agency (IEA); European Environmental Agency (EEA).

CO_2 emissions in 2002

Tonnes per person

see also graph on page 83

- United States — 20
- Australia — 18
- Canada — 14
- European Union — 12
- Norway
- Russia
- Japan
- South Africa
- France — 6
- China
- Brazil
- India

High-income countries

Global average

Low-income countries

Authors of the texts

Emmanuelle Bournay
HAZARDOUS WASTE MANAGEMENT
Waste, recyclers and recycled
geographer and cartographer; joint author of
Vital Waste Graphics, UNEP – Basel Convention, Geneva, 2004.

Any Bourrier
WEAPONS OF MASS DESTRUCTION
Weapons for rich ... and poor
journalist at Radio France Internationale.

Philippe Bovet
INDUSTRIAL POLLUTION
Who causes industrial accidents?
freelance journalist

François Carré
FISHING
Ocean resources under threat
professor at Paris-Sorbonne university, chair of
the Sea and Seashore committee at France's
National Geographical Commission (CNFG),
joint author of *Milieux littoraux. Nouvelles perspectives d'étude*, L'Harmattan, 2005.

Benjamin Dessus
RENEWABLE ENERGY
Renewable energy, fact and fiction
chair of Global Chance, author, with Hélène
Gassin, of *So Watt? L'énergie, une affaire de
citoyens*, Editions de l'Aube, 2004.

Frédéric Durand
POLAR REGIONS
Polar ice caps melting faster
CLIMATE CHANGE
Point of no return for global warming
lecturer at Toulouse-Le Mirail university
author of *La Jungle, la Nation et le Marché.
Chronique indonésienne*, L'Atalante, 2001.

Marc Laimé
WATER MANAGEMENT
Water becoming a rarity
journalist, author of *Le dossier de l'eau. Pénurie, pollution, corruption*, Seuil, 2003.

Damien Millet
RAW MATERIALS
The South depends on its exports
FOOD
Losing the battle against hunger
chair of the French Committee for the Abolition of Third World Debt (CADTM), author
of *L'Afrique sans dette*, CADTM/Syllepse,
2005.

Thierry Paquot
URBANISATION
Urban development trends
lecturer in town planning at Paris-XII university, author of *Demeure terrestre. Enquête
vagabonde sur l'habiter*, L'Imprimeur, Paris,
2005.

Philippe Rivière
PUBLIC HEALTH
Widening health care gap
journalist at *Le Monde diplomatique*

Mycle Schneider
NUCLEAR POWER
Nuclear power for civilian and military use
journalist and scientific expert, director of
the World Information Service on Energy
(Wise-Paris)

Michel Urvoy
GM ORGANISMS
GM organisms, too much, too soon
journalist, head of the economic and farming
pages of the daily Ouest-France (Rennes)